ACTUARIAL SCIENCE

THE UNIVERSITY OF WESTERN ONTARIO SERIES IN PHILOSOPHY OF SCIENCE

A SERIES OF BOOKS
IN PHILOSOPHY OF SCIENCE, METHODOLOGY,
EPISTEMOLOGY, LOGIC, HISTORY OF SCIENCE
AND RELATED FIELDS

VOLUME 39

ADVANCES IN THE STATISTICAL SCIENCES

Festschrift in Honor of Professor V. M. Joshi's 70th Birthday

VOLUME VI

ACTUARIAL SCIENCE

Edited by

IAN B. MacNEILL and GARY J. UMPHREY

Department of Statistical and Actuarial Sciences,
The University of Western Ontario

Associate editors:

BEDA S. C. CHAN

and

SERGE B. PROVOST

Department of Statistical and Actuarial Sciences,
The University of Western Ontario

D. REIDEL PUBLISHING COMPANY

A MEMBER OF THE KLUWER ACADEMIC PUBLISHERS GROUP

DORDRECHT / BOSTON / LANCASTER / TOKYO

Library of Congress Cataloging in Publication Data

Actuarial science.

(Advances in the statistical sciences; v. 6)
(The University of Western Ontario series in philosophy of science; v. 39)
1. Insurance—Mathematics—Congresses. 2. Actuaries—Congresses. I. MacNeill, Ian B., 1931– . II. Umphrey, Gary J., 1953– . III. Series. IV. Series: University of Western Ontario series in philosophy of science; v. 39.
QA276.A1A39 vol. 6 519.5 s 86–29678
[HG8782] [368'.01]

ISBN-13: 978-94-010-8627-1 e-ISBN-13: 978-94-009-4796-2
DOI: 10.1007/978-94-009-4796-2

Published by D. Reidel Publishing Company,
P.O. Box 17, 3300 AA Dordrecht, Holland.

Sold and distributed in the U.S.A. and Canada
by Kluwer Academic Publishers,
101 Philip Drive, Assinippi Park, Norwell, MA 02061, U.S.A.

In all other countries, sold and distributed
by Kluwer Academic Publishers Group,
P.O. Box 322, 3300 AH Dordrecht, Holland.

TABLE OF CONTENTS

Contents of the Other Volumes of the Joshi Festschrift vii

Preface xiii

Introduction to Volume VI xv

JAMES C. HICKMAN
Connections Between Graduation and Forecasting 1

JAMES D. BROFFITT
Isotonic Bayesian Graduation with an Additive Prior 19

ESTHER PORTNOY
Bootstrapping a Graduation 41

HANS U. GERBER
Actuarial Applications of Utility Functions 53

N. U. PRABHU
A Class of Ruin Problems 63

S. DAVID PROMISLOW
Comparing Risks 79

STUART KLUGMAN
Inference in the Hierarchical Credibility Model 95

HARRY H. PANJER
Models of Claim Frequency 115

R. M. DUMMER
Analyzing Casualty Insurance Claim Counts 127

ELIAS S. W. SHIU
Immunization—The Matching of Assets and Liabilities . . . 145

JOHN A. BEEKMAN
Ornstein-Uhlenbeck Stochastic Processes Applied to Immunization 157

S. BROVERMAN
A Note on Variable Interest Rate Loans 165

PATRICK L. BROCKETT AND NAIM SIPRA
Linearity and Gaussianity of Interest Rate Data: An Empirical Time Series Test 173

PHELIM P. BOYLE
Perspectives on Mortgage Default Insurance 185

KEITH P. SHARP
Time Series Analysis of Mortgage Rate Insurance 201

SAMUEL H. COX, JR., AND CHENG-KUN KUO
Underwriting Traders of Financial Futures 219

D. S. RUDD
Mergers of Life Companies and the Blurring of Boundaries Among Financial Institutions—Effects on the Actuarial Profession 231

K. W. STEWART
Commentary on Rudd's Talk 235

GEORGE DINNEY
The Search for New Forms of Life 239

CONTENTS OF THE OTHER VOLUMES OF THE JOSHI FESTSCHRIFT

VOLUME I

Applied Probability, Stochastic Processes, and Sampling Theory

W. J. ANDERSON / Probabilistic Models of the Photographic Process

D. BINDER, J. KOVAR, S. KUMAR, D. PATON & A. VAN BAAREN / Analytic Uses of Survey Data: A Review

M. BLAIS / Transience in a Queueing System with a Finite Number of Locally Interacting Servers

D. R. BRILLINGER / Fitting Cosines: Some Procedures and Some Physical Examples

A. CORNISH / V. M. Joshi and the Markov Oscillation Problem

A. F. DESMOND / An Application of Some Curve-Crossing Results for Stationary Stochastic Processes to Stochastic Modelling of Metal Fatigue

R. FERLAND & G. GIROUX / The Convergence of the Solution of a Boltzmann Type Equation Related to Quantum Mechanics

W. A. FULLER / Estimators of the Factor Model for Survey Data

J. GANI / Some Recent Work in Epidemic Models

M. GHOSH / On Admissibility and Uniform Admissibility in Finite Population Sampling

M. A. HIDIROGLOU & D. G. PATON / Some Experiences in Computing Estimates and Their Variances Using Data from Complex Survey Designs

R. J. KULPERGER / Central Limit Theorems for Cluster Point Processes

D. E. MATTHEWS, CH. E. MINDER & I. McMILLAN / A Stochastic Model for the Effect of Incident Light Intensity on CO_2 Uptake in Leaves

D. L. McLEISH & C. G. SMALL / Likelihood Asymptotics for the Discrimination Problem

T. T. NGUYEN / On Fréchet Bounds of Bivariate Distributions

B. L. S. PRAKASA RAO / Asymptotic Theory of Estimation in Nonlinear Regression

C. M. RAMSAY / Strong Limit Theorems for Sums of Random Variables Defined on a Finite Markov Chain

R. M. ROYALL / Overlooked Correlation in Finite Population Inference

A. R. SEN & P. K. SEN / Estimation of the Characteristics of Rare Animals Based on Inverse Sampling at the Second Occasion

M. E. THOMPSON / Ideas from the Foundations of Sampling Applied to the One-Way Layout

P. TODOROVIC / Limit Theorems Arising in Soil Erosion Modelling

S. L. WARNER / Identifying Rational Opinion-Formation with the Overlapping Information Model

VOLUME II

Foundations of Statistical Inference

M. AKAHIRA & K. TAKEUCHI / On the Definition of Asymptotic Expectation

M. A. ALI / Missing Value Problems in Multiple Linear Regression with Two Independent Variables

M. M. ALI / A Bound for the Tail Area of the t Distribution for Samples from a Symmetrically Truncated Normal Population

C. R. BLYTH & J. V. BONDAR / A Neyman-Pearson-Wald View of Fiducial Probability

J. V. BONDAR / How Much Improvement Can a Shrinkage Estimator Give?

A. P. DEMPSTER / Probability and the Future of Statistics

M. EVANS, D. A. S. FRASER & G. MONETTE / Statistical Principles and Tangent Models

D. A. S. FRASER & H. MASSAM / An Algorithm for Concave Regression

V. P. GODAMBE / Data Based Choice of an Ancillary Statistic

I. GUTTMAN & M. S. SRIVASTAVA / Bayesian Method of Detecting Change Point in Regression and Growth Curve Models

M. S. HAQ / On the Prediction of the Difference Between Responses from Two Linear Models

L. V. HEDGES & I. OLKIN / Statistical Inference for the Overlap Hypothesis

H. E. KYBURG, JR. / The Basic Bayesian Blunder

C.-I. C. LEE / Maximum Likelihood Estimates for Stochastically Ordered Multinomial Populations with Fixed and Random Zeros

D. V. LINDLEY / Bernoulli Pairs with Invariant Reversals: An Example of Partial Likelihood

S. NISHISATO / Robust Techniques for Quantifying Categorical Data

A. PLANTE / A Decision-Likelihood Solution to the Problem of Comparing Two Simple Hypotheses

J. L. POLLOCK / Sketch of the Theory of Nomic Probability

S. B. PROVOST / Testing for the Nullity of the Multiple Correlation Coefficient with Incomplete Multivariate Data

A. K. Md. E. SALEH & P. K. SEN / On Shrinkage and Preliminary Test M-Estimation in a Parallelism Problem

T. SEIDENFELD / Entropy and Uncertainty

B. SKYRMS / Dynamic Coherence

D. S. TRACY & K. G. JINADASA / On Ultrastructural Relationships Models

VOLUME III

Time Series and Econometric Modelling

B. ABRAHAM / Outliers in Time Series

H. AKAIKE / Some Reflections on the Modelling of Time Series

L. A. AROIAN / Recent Results for Time Series in M Dimensions

E. B. DAGUM / Monthly versus Annual Revisions of Concurrent Seasonally Adjusted Series

J.-M. DUFOUR / Linear Wald Methods for Inference on Covariances and Weak Exogeneity Tests in Structural Equations

Q. P. DUONG / Model Selection and Forecasting: A Semi-Automatic Approach

A. FEUERVERGER / On Some ECF Procedures for Testing Independence

C. W. J. GRANGER / Are Economic Variables Really Integrated of Order One?

E. J. HANNAN / Approximation of Linear Systems

O. G. JENSEN & L. MANSINHA / Excitation of Geophysical Systems with Fractal Flicker Noise

B. KEDEM / A Fast Graphical Goodness of Fit Test for Time Series Models

T. S. KHEOH & A. I. McLEOD / On the Efficiency of a Strongly Consistent Estimator in ARMA Models

E. MAASOUMI / The Approximate Moments of the 3SLS Reduced Form Estimator and a MELO Combination of OLS-3SLS for Prediction

T. A. PETERS / The Finite Sample Moments of OLS in Dynamic Models When Disturbances are Small

P. C. B. PHILLIPS / Fractional Matrix Calculus and the Distribution of Multivariate Tests

S. POWER / Asymptotic Properties of Single Equation Errors in Variables Estimators in Rational Expectations Models

R. S. SINGH, A. ULLAH & R. A. L. CARTER / Nonparametric Inference in Econometrics: New Applications

D. S. STOFFER & T. PANCHALINGAM / A Walsh-Fourier Approach to the Analysis of Binary Time Series

B. C. SUTRADHAR, I. B. MacNEILL & H. F. SAHRMANN / Time Series Valued Experimental Designs: One-Way Analysis of Variance with Autocorrelated Errors

T. TERÄSVIRTA / Smoothness in Regression: Asymptotic Considerations

H. TSURUMI / Use of the Mean Squared Errors of Forecasts in Testing for Structural Shift: A Comparison with the Chow Test for an Undersized Case

Y. VAN HUI & W. K. LI / Predicting Demands in a Multi-Item Environment

M. R. VEALL / Bootstrapping and Forecast Uncertainty: A Monte Carlo Analysis

H. D. VINOD / Confidence Intervals for Ridge Regression Parameters

V. ZINDE-WALSH & A. ULLAH / On Robustness of Tests of Linear Restrictions in Regression Models with Elliptical Error Distributions

VOLUME IV

Stochastic Hydrology

B. A. BODO AND T. E. UNNY / On the Outputs of the Stochasticized Nash-Dooge Linear Reservoir Cascade

F. CAMACHO, A. I. McLEOD & K. W. HIPEL / The Use and Abuse of Multivariate Time Series Models in Hydrology

J. P. CHANUT, M. I. EL-SABH, M. MARCHETERRE & R. ROY / A Stochastic Modelling of Tidal Current Measurements

N. R. DALEZIOS, P. A. TYRASKIS & B. G. LATHAM / Autoregressive Empirical Modelling of Multiple Precipitation Time Series

J. KELMAN / Statistical Approach to Floods

C. LABATIUK & K. ADAMOWSKI / Application of Nonparametric Density Estimation to Computation of Flood Magnitude/Frequency

D. K. PICKARD & E. M. TORY / A Markov Model for Sedimentation: Fundamental Issues and Insights

S. E. SERRANO, & T. E. UNNY / Stochastic Partial Differential Equations in Hydrology

M. A. STEPHENS / Tests for the Extreme-Value and Weibull Distributions: Some Recent Developments

R. M. THOMPSTONE, K. W. HIPEL & A. I. McLEOD / Simulation of Monthly Hydrological Time Series

T. E. UNNY / Solutions to Nonlinear Stochastic Differential Equations in Catchment Modelling

S. YAKOWITZ & M. KARLSSON / Nearest Neighbor Methods for Time Series, with Application to Rainfall/Runoff Prediction

VOLUME V

Biostatistics

S. B. BULL & A. DONNER / Derivation of Large Sample Efficiency of Multinomial Logistic Regression Compared to Multiple Group Discriminant Analysis

J. W. CHAPMAN, J. ETEZADI-AMOLI, P. J. SELBY, N. F. BOYD & D. DALLEY / Statistical Ramifications of Linear Analogue Scales in Assessing the Quality of Life of Cancer Patients

A. CIAMPI, C.-H. CHANG, S. HOGG & S. McKINNEY / Recursive Partition: A Versatile Method for Exploratory Data Analysis in Biostatistics

A. DONALD & A. DONNER / The Effect of Clustering on the Analysis of Sets of 2 × 2 Contingency Tables

W. T. FEDERER & B. R. MURTY / Uses, Limitations, and Requirements of Multivariate Analyses for Intercropping Experiments

W. W. HAUCK / Estimation of a Common Odds Ratio

S. KONISHI & A. K. GUPTA / Inferences about Interclass and Intraclass Correlations from Familial Data

J. KOVAL & A. DONNER / Estimation Under the Correlated Logistic Model

D. KREWSKI, R. T. SMYTHE & D. COLIN / Tests for Trend in Binomial Proportions with Historical Controls: A Proposed Two-Stage Procedure

J. F. LAWLESS & K. SINGHAL / Regression Methods and the Exploration of Large Medical Data Bases

N. MANTEL & S. R. PAUL / Goodness-of-Fit Issues in Toxicological Experiments Involving Litters of Varying Size

M. M. SHOUKRI & P. C. CONSUL / Some Chance Mechanisms Generating the Generalized Poisson Probability Models

R. VIVEROS & D. A. SPROTT / Maximum Likelihood Estimation in Quantal Response Bioassay

M. N. WALSH, J. J. HUBERT & E. M. CARTER / Estimation Methods for Symmetric Parabolic Bioassays

S. D. WALTER / Point Estimation of the Odds Ratio in Sparse 2 × 2 Contingency Tables

G. A. WELLS & A. DONNER / Development of Formulas for the Bias and Mean Square Error of the Logit Estimator

PREFACE

On May 27–31, 1985, a series of symposia was held at The University of Western Ontario, London, Canada, to celebrate the 70th birthday of Professor V. M. Joshi. These symposia were chosen to reflect Professor Joshi's research interests as well as areas of expertise in statistical science among faculty in the Departments of Statistical and Actuarial Sciences, Economics, Epidemiology and Biostatistics, and Philosophy.

From these symposia, the six volumes which comprise the "Joshi Festschrift" have arisen. The 117 articles in this work reflect the broad interests and high quality of research of those who attended our conference. We would like to thank all of the contributors for their superb cooperation in helping us to complete this project.

Our deepest gratitude must go to the three people who have spent so much of their time in the past year typing these volumes: Jackie Bell, Lise Constant, and Sandy Tamowski. This work has been printed from "camera ready" copy produced by our Vax 785 computer and QMS Lasergraphix printers, using the text processing software TEX. At the initiation of this project, we were neophytes in the use of this system. Thank you, Jackie, Lise, and Sandy, for having the persistence and dedication needed to complete this undertaking.

We would also like to thank Maria Hlawka-Lavdas, our systems analyst, for her aid in the layout design of the papers and for resolving the many difficult technical problems which were encountered. Nancy Nuzum and Elly Pakalnis have also provided much needed aid in the conference arrangements and in handling the correspondence for the Festschrift.

Professor Robert Butts, the Managing Editor of *The University of Western Ontario Series in Philosophy of Science* has provided us with his advice and encouragement. We are confident that the high calibre of the papers in these volumes justifies his faith in our project.

In a Festschrift of this size, a large number of referees were needed. Rather than trying to list all of the individuals involved, we will simply say "thank you" to the many people who undertook this very necessary task for us. Your contributions are greatly appreciated.

Financial support for the symposia and Festschrift was provided by The University of Western Ontario Foundation, Inc., The University of Western Ontario and its Faculties of Arts, Science, and Social Science, The UWO Statistical Laboratory, and a conference grant from the Natural Sciences and Engineering Research Council of Canada. Their support is gratefully acknowledged.

Finally, we would like to thank Professor Joshi for allowing us to hold the conference and produce this Festschrift in his honor. Professor Joshi is a very modest man who has never sought the limelight. However, his substantial contributions to statistics merit notice (see Volume I for a bibliography of his papers and a very spiffy photo). We hope he will accept this as a tribute to a man of the highest integrity.

INTRODUCTION TO VOLUME VI

In 1963 the British astronomer Edmond Halley, whose name is memorialized in the comet that recently left the vicinity of our planet, presented a paper to the Royal Society entitled, "An estimate of the degrees of mortality of mankind drawn from curious tables of the births and funerals of the city of Breslaw; with an attempt to ascertain the price of annuities upon lives". Since then, the actuarial literature has benefitted from many of the numerous advances made in the fields of numerical analysis and mathematical statistics.

The development of the actuarial profession in a variety of fields parallels an ever-increasing utilization of statistical techniques to analyze actuarial concepts once they have been translated into probabilistic terms. This is evidenced by most of the nineteen articles comprising this volume of the Festschrift in honor of Professor Joshi's 70th birthday.

The first three papers of this volume deal with the popular research topic of graduation. The foundations of graduation and the connections between some graduation techniques and concepts used in forecasting are discussed by Hickman. Broffitt uses an isotonic Bayesian approach with an additive prior for the graduation of mortality rates. Portnoy applies the bootstrap idea to graduation.

Risk aversion has been a constant concern in the insurance world. This is reflected in the next six papers which are all related to risk analysis. Gerber discusses some actuarial applications of utility functions including reinsurance. Prahbu studies a class of ruin problems using compound renewal processes, while Promislow considers the problem of comparing probability distributions by their degree of risk. Klugman obtains variance estimates of point estimators in order to make some inferences about the hierarchical credibility model wherein the various classes of insureds are partitioned into several groups. Panjer is concerned with modelling the number of claims that arise from a portfolio of insured risks, while Dummer proposes a statistical model for analyzing casualty insurance claim counts when delays in claim reporting are experienced.

The next seven papers are indicative of the actuarial profession's growing interest in the investment world. The theory of immunization is the focus of the papers by Shiu and Beekman; the former gives a brief review of this theory and discusses possible extensions using some inequalities of convex functions, while the latter enriches this theory by modelling the instantaneous borrowing and lending rate on the Ornstein-Uhlenbeck stochastic process. Broverman presents a method of determining the payment for a variable interest rate loan and explores its effects on the amortization of a debt. Brockett and Sipra use a time series procedure in order to test for the linearity and the Gaussianity of interest rate data. Boyle investigates alternative approaches to the pricing and risk management of mortgage default insurance, arguing that traditional approaches are inadequate, while Sharp shows that a particular time series model which satisfactorily represents monthly interest rate data can be used in the determination of mortgage rate insurance premiums. Cox and Kuo propose a framework for the analysis of cash requirements for traders of financial futures and address the problem of finding an optimal procedure for trading futures contracts indefinitely.

The remaining articles constitute accounts of the addresses given at the conference by three practicing actuaries. Rudd discusses the effects of the future demands for and on the actuarial profession of mergers of life companies and of the blurring of boundaries among financial institutions. Stewart provides a commentary on Rudd's address, considers some expectations concerning the evolution of financial services and draws some lessons for the actuarial profession. Dinney discusses the concept of Universal Life and its future in the life insurance industry.

We are grateful to the authors for the quality of their contributions. They have proficiently tackled a variety of current problems arising in actuarial science. Their analyses, results and conclusions should prove to be of substantial value to the actuarial profession.

James C. Hickman [1]

CONNECTIONS BETWEEN GRADUATION AND FORECASTING

ABSTRACT

The motivation for the development of graduation methods is usually based on one of two foundations:

1. Mathematical programming. An objective function is specified and the set of graduated values is found so that an optimal value of the objective function is attained.
2. Statistical analysis. A statistical model is specified and the set of graduated values that attain a goal stated in statistical terms such as minimum variance or a posterior mean is found.

Moving-weighted-average, Whittaker and parametric graduation methods have well-known twin foundations. There are connections between moving-weighted-average methods and discounted least squares and time series analysis methods used in forecasting. The common model selection tools (examination of sample autocorrelations, and checks for shifts in variance) used in forecasting have not entered graduation. Illustrations of the implications of the use of model selection tools are presented.

1. FOUNDATIONS OF GRADUATION

Andrews and Nesbitt (1965) defined graduation as "an effort to represent a physical phenomenon by a systematic revision of some observations of the phenomenon". From this definition it is clear that graduation pervades science.

Within actuarial science the principal application of graduation has been in representing human mortality in the form of a life table based on observa-

---

[1] School of Business, University of Wisconsin, Madison, Wisconsin 53706

I. B. MacNeill and G. J. Umphrey (eds.), Actuarial Science, 1–17.

tions of q_x. However, systematic revisions of observations occur in numerous other actuarial applications. These include disability rates, disability recovery rates, withdrawal rates, rates of investment return, and loss ratios.

Actuarial literature is rich with contributions to the theory of graduation. Because of the links between actuarial graduation and other scientific smoothing methods, there are more connections between actuarial science and other sciences on this topic than with most of the topics that constitute actuarial science.

Contributions to graduation can be classified according to whether they build on a *statistical* or a *mathematical programming* foundation. A statistical foundation is one that adopts a model with an explicit probability component. Contributors who adopt a statistical foundation have usually stated their goals in statistical terms. That is, the goal may be to produce a minimum variance unbiased estimate, within the framework of the selected model, or to find a posterior mean that combines data and prior information. Those who build on a mathematical programming foundation select an objective function, influenced by their view of the purpose of the graduation and the nature of the observations, and find an extreme value of this function.

In order to illustrate the twin foundations of graduation, we will organize some recent contributions to the theory according to this division. We start with some contributions that appear to be based on a mathematical programming foundation.

Schuette (1978), Chan *et al.* (1982) and Chan *et al.* (1985) developed some aspects of the Whittaker formulation of the goal of graduation. In this formulation the goal of graduation is to find a vector v which will minimize the loss function

$$F(v) + hS(v). \tag{1.1}$$

These authors used norms other than the traditional ℓ_2 norm in $F(v)$, the measure of lack of fit, and in $S(v)$, the measure of the lack of smoothness. The study note by Greville (1973), from which actuarial students for over a decade learned the fundamentals of graduation, essentially takes a mathematical programming view of graduation. Schoenberg (1964) pointed out connections between Whittaker's method and spline function approximations. Shiu (1985) used the methods of mathematical programming to derive formulas for the coefficients of minimum R_z moving-weighted-average (MWA) formulas.

This paper will concentrate on the models and methods used by contributors to the theory of graduation who have built on a statistical foundation. Therefore, our organization of the recent contributions to this line of research will be more complete and have more structure.

A. Fitting Distributions

1. Bayesian methods. Kimeldorf and Jones (1967) used conjugate analysis with multinormal prior, likelihood and posterior distributions. Hickman (1967), for observations of a different type, used a multinomial likelihood with conjugate Dirichlet prior and posterior distributions.
2. Least squares. Tenenbein and Vanderhoof (1980) used weighted least squares and transformed observations to fit an expanded class of Gompertz functions to mortality data.
3. Maximum likelihood. Chan and Panjer (1983) applied the maximum likelihood estimation principle to a wide variety of mortality distributions, including the traditional Makeham and Gompertz distributions, and to various ways of organizing observations.
4. Moments. At the present time basic actuarial education in North America does not stress the method of moments in connection with graduation. However, Henderson (1938) devoted an entire chapter to the method of moments.
5. Special methods. Hardy's method of estimating the Makeham constant c, as taught by Miller (1946), is an example of a special method of estimating a parameter of a distribution that is difficult to classify according to the usual estimation methods.

B. Smoothing—with probability elements but without fitting a distribution.

1. MWA. Clearly one of the two rationales for this graduation method rests on a statistical model. That is, each observation consists of a value determined by a local polynomial plus a random error term that satisfies the usual constant variance and uncorrelated assumptions of linear statistical models. Sheppard (1914) and Borgan (1979) discussed MWA methods by building on a statistical foundation.
2. Whittaker. One of Whittaker's (1924) descriptions of the graduation method that bears his name had an explicit probability component. However, in the graduation method only the mode of the posterior distribution of parameters was of concern. The work of Craven and Wahba (1979) involves smoothing noisy data with an explicit random error term and with the minimization of a loss function like (1.1), except that the roughness measures involves the integration of the squared derivatives of the fitted function.

To crystalize the distinction being made, one needs only to consider two traditional graduation methods. As indicated above, Whittaker supplied two rather different motivations for his methods. In his 1923 paper a mathematical programming foundation is used. The 1924 book with Robinson provides a Bayesian statistical foundation. In the current graduation study note by London (1984), a reduction of variance argument for MWA methods is provided in Section 3.4. In Section 3.5, the same method is motivated by a programming argument.

2. FORECASTING

The purpose of this paper is to examine two traditional graduation methods from the viewpoint of applied statistics. The goal is modest. No new methods will be proposed. Instead, modification of two existing methods will be suggested to bring the methods in closer correspondence with the statistical foundations on which they can be built. Hoem (1984) had a much more ambitious goal. Several of his insightful comments about linear graduation methods are in the spirit of this paper. Hoem says: "For this purpose, general linear graduation is introduced in Section 3 and a method is provided which is optimal under mathematical restrictions. Such restrictions frequently will be violated in practical applications, but moving average methods have the nice feature of being operative even when the restrictions are satisfied only as a crude approximation. This robustness property,..., may be one of the strongest motivations for applying moving averages."

The term "forecasting" rather than "applied statistics" was used in the title. Applied statistics includes the design and analysis of experiments as well as the analysis of observations from undesigned or observational studies. In designed experiments it is frequently possible to manipulate the design to bring the observations into closer correspondence to the model by which they will be analyzed. In observational studies, including forecasting, the data must help select a model; a model is not a *priori* and uncritically chosen. Formal tests of hypotheses are inappropriate but informal tests of significance are used in model selection and criticism. Graduation is more closely related to observational studies, and particularly to forecasting, than to the design of experiments, a topic in applied statistics.

The usual model in forecasting is

$$\text{Response}_t = \text{Fit}_t + \text{Error}_t, \tag{2.1}$$

where future responses are to be forecast and Fit_t is a function of earlier values of the response variables or a function of other explanatory variables which may be observed. If Fit_t is a linear function in the parameters, the usual time series and regression models emerge. The error terms, to be denoted by e_t, are assumed to be such that:

$$e_t \text{ is } N(0,\sigma^2), \text{ for all } t, \text{ and } E[e_t e_s] = 0, \text{ for } t \neq s. \tag{2.2}$$

The paradigm of the forecasting branch of applied statistics calls for the iteration of a three step process.

1. Identification of a model by use of plots of the response variable and its differences, and of various candidate explanatory variables. These plots are augmented by descriptive statistics such as correlation and autocorrelation coefficients.
2. Estimation of the parameters of the selected model. This is essentially a computational task.
3. Criticism of the model using

$$\text{Residual}_t = \text{ Observed Response}_t - \text{ Fit}_t. \tag{2.3}$$

   If it is plausible to believe that the residuals behave in accord with standard assumptions (2.2), the analysis can proceed to forecasting. If it is not plausible, then revised identification is indicated. This step may involve transforming the data, adding or deleting a parameter or selecting a model from a different class.

The idea is to continue this process until the residuals do not appear to contain useful information.

3. GRADUATION AND FORECASTING

Because of the absence of iteration in the process, graduation methods have not followed the paradigm of applied statistics. We will examine the MWA and Whittaker methods with the goal of modifying them to bring them into accord with assumptions (2.2).

a. MWA. In this method each graduated value is determined from

$$v_x = \sum_{r=-m}^{m} a_r u_{x+r}, \tag{3.1}$$

where u denotes observed values, v graduated values and a weights. Usually the weights have been constrained so that if the observations u_{x+r}, $r = -m, \ldots, m$, are on a polynomial of degree k, then $v_x = u_x$. The degree is usually $k = 3$.

This use of a local polynomial as a fit function has an analogue in forecasting. Discounted least squares is a method for reducing the influence of remote observations in estimating the fit function. The topic is explained by Abrahams and Ledolter (1983). In this approach we let

$$P_k(x) = \sum_{i=0}^{k} \beta_i x^i. \tag{3.2}$$

Then the parameters of the polynomial fit function for a MWA graduation, would be determined by finding a minimum value of

$$S(\beta) = \sum_{r=-m}^{m} w_r \left[u_{x+r} - P_3(x+r)\right]^2 , \tag{3.3}$$

where $0 \leq w_r \leq 1$. In actuarial graduations, $w_r = 1$ and the MWA weights are the coefficients of the values of u_{x+r}, $r = -m, \ldots, m$, in the estimate of β_0.

The use of local polynomials as a fit function is well known in applied statistics. However, the use of least squares as an estimation technique without checking the residuals given in (2.3) for autocorrelations and shifts in variance is subject to criticism.

b. Whittaker. The minimization of the loss function (1.1) has as a prerequisite the specification of F and S. The fit term is usually taken as

$$\sum_{1}^{n} w_x \left(u_x - v_x\right)^2 \tag{3.4}$$

and, with the statistical motivation, the weights w_x are taken to be proportional to $\text{Var}(U_x)^{-1}$. Applied statisticians would recognize this as a weighted least squares approach. However, the failure to check the data to see if the observations, $U_x, x = 1, 2, \ldots, n$, are autocorrelated would be criticized.

Likewise, the selection of a smoothness criterion without reference to the data would not be in the spirit of applied statistics. With mortality data it is typical for

$$S = \sum_{1}^{n-z} \left(\Delta^z v_x\right)^2$$

or

$$S = \sum_{1}^{n-2} \left(\Delta v_{x+1} - (1+r)\Delta v_x\right)^2 \tag{3.5}$$

and z, the order of differencing as well as the alternative constant r defining the exponential function to which the fit function is constrained, are usually determined by convention or prior experience.

4. MORTALITY DATA

Are the canonical assumptions of applied statistics (2.2) realized in mortality data? If the answer is no, corrective measures such as transformations will have to be employed before we can expect satisfactory results in graduation by the standards of applied statistics.

a. Constant variance. Since

$$\text{Var}\ [\theta_x/E_x] \doteq q_x p_x/E_x, \tag{4.1}$$

it is clear that the constant variance assumption is violated. Some sort of transformation or weighted least squares must be used if satisfactory results are to be expected. Hickman and Miller (1977) discussed transformations within the context of Bayesian graduation. London (1981) discussed a version of weighted least squares in MWA methods. The scope of applicability of transformations and weighted least squares can be expanded beyond the two graduation methods discussed in these two papers.

b. Uncorrelated observations. Consider the following simple Lexis diagram which summarizes the data available from a small mortality study involving two ages and two years of exposure. It will be assumed that migration is zero.

		Year 1	Year 2
Age	1	θ_{11}, E_{11}	θ_{12}, E_{12}
	2	θ_{21}, E_{21}	θ_{22}, E_{22}

It is natural to assume, in the absence of an epidemic or an instantaneous introduction of improved medical technology, that

$$\begin{aligned} &\text{Cov}\left[\frac{\theta_{11}}{E_{11}}, \frac{\theta_{21}}{E_{21}}\right] = \text{Cov}\left[\frac{\theta_{12}}{E_{12}}, \frac{\theta_{22}}{E_{22}}\right] \\ =\ &\text{Cov}\left[\frac{\theta_{11}}{E_{11}}, \frac{\theta_{21}}{E_{21}}\right] = \text{Cov}\left[\frac{\theta_{12}}{E_{12}}, \frac{\theta_{22}}{E_{22}}\right] \\ =\ &\text{Cov}\left[\frac{\theta_{21}}{E_{21}}, \frac{\theta_{22}}{E_{22}}\right] = 0. \end{aligned} \tag{4.2}$$

However, we would have

$$\operatorname{Cov}\left[\frac{\theta_{11}}{E_{11}}, \frac{\theta_{22}}{E_{22}}\right] < 0$$

because of the multinomial distribution of $(\theta_{11}, \theta_{22}, E_{11} - \theta_{11} - \theta_{22})$.

For most mortality studies, the issue is to find

$$\operatorname{Cov}\left[\frac{\theta_{11} + \theta_{12}}{E_{11} + E_{12}}, \frac{\theta_{21} + \theta_{22}}{E_{21} + E_{11} - \theta_{11}}\right]. \tag{4.3}$$

This can be done by a succession of conditional expectations. If we assume that E_{11}, E_{12}, E_{21} are constants, and that the θ's are random variables, then

$$E\left[\frac{\theta_{11} + \theta_{12}}{E_{11} + E_{12}}\right] = \frac{(E_{11} + E_{12})\, q_1}{E_{11} + E_{12}} = q_1$$

and

$$\begin{aligned} &E_{\theta_{11}} E_{\theta_{21}, \theta_{22} | \theta_{11}} \left[\frac{\theta_{21} + \theta_{22}}{E_{21} + E_{11} - \theta_{11}}\right] \\ &= E_{\theta_{11}} \left[\frac{(E_{21} + E_{11} - \theta_{11})\, q_2}{E_{21} + E_{11} - \theta_{11}}\right] = q_2. \end{aligned} \tag{4.4}$$

By (4.4) we have established unbiasedness. Then the covariance (4.3) can be computed by evaluating

$$\begin{aligned} &E_{\theta_{11}} \left\{E_{\theta_{12}, \theta_{21}, \theta_{22} | \theta_{11}} \left[\left(\frac{\theta_{11} + \theta_{12}}{E_{11} + E_{12}}\right)\left(\frac{\theta_{21} + \theta_{22}}{E_{21} + E_{11} - \theta_{11}}\right)\right]\right\} \\ =&E_{\theta_{11}} \left[\frac{\theta_{11}\,(E_{21} + E_{11} - \theta_{11})\, q_2 + (E_{12})\,(q_1)\,(E_{21} + E_{11} - \theta_{11})\, q_2}{(E_{11} + E_{12})\,(E_{21} + E_{11} - \theta_{11})}\right] \\ =&E_{\theta_{11}} \left[\frac{\theta_{11} q_2 + E_{12} q_1}{E_{11} + E_{12}}\right] = \frac{E_{11} q_1 q_2 + E_{12} q_1 q_2}{E_{11} + E_{12}} = q_1 q_2, \end{aligned}$$

and the covariance in (4.3) is zero.

Consequently, actuaries can expect their usual estimates of q_x to be uncorrelated. However, the plausible correlation assumptions (4.2) are often not realized in observations because of common factors which may create correlations and thereby invalidate the canonical assumptions of applied statistics.

5. MODIFYING GRADUATION METHODS

Clearly the steps to modify traditional graduation methods to accommodate deviations from assumptions (2.2) will require iteration. These steps will include stabilizing the variance of observations, selecting the degree of the local polynomial and adjusting the graduation method to account for correlated error terms.

To illustrate, we will follow Hickman and Miller (1977). If $U_x = \theta_x / E_x$, then Var [arcsin $\sqrt{U_x}$] is approximately $(4E_x)^{-1}$. Then working with

$$Y_x = \sqrt{E_x} \arcsin \sqrt{U_x},$$

we can expect that $\mathrm{Var}(Y_x)$, $x = 1, 2 \ldots, n$, are approximately constant. We can then develop an informal test about the degree of a polynomial fit function. Assume

$$Y_x = P_h(x) + e_x.$$

Then

$$\Delta^\ell Y_x = \Delta^\ell e_x, \qquad \ell \geq h+1,$$

$$\mathrm{Var}\left(\Delta^\ell Y_x\right) = \binom{2\ell}{\ell}\sigma^2,$$

$\Delta^\ell Y_x$ has a $N\left(0, \binom{2\ell}{\ell}\sigma^2\right)$ distribution, and

$$\left(\Delta^\ell Y_x\right)^2 / \mathrm{Var}\left(\Delta^\ell Y_x\right)$$

has a chi-squared distribution with one degree of freedom.
Then

$$\sum_{x \in N} \frac{\left(\Delta^{\ell'} Y_x\right)^2}{\binom{2\ell'}{\ell'}\sigma^2}$$

has a chi-squared distribution with n degrees of freedom, where n is the number of terms in the index set N, so long as the index set is selected so that individual Y_x terms do not appear in more than one term of the summation. Then if ℓ and ℓ' are greater than h,

$$F = \frac{\sum_{x \in N} \left(\Delta^\ell Y_x\right)^2 / \left[n \binom{2\ell}{\ell}\right]}{\sum_{x \in D} \left(\Delta^{\ell'} Y_x\right)^2 / \left[d \binom{2\ell'}{\ell'}\right]} \tag{5.1}$$

will have an F distribution with n and d degrees of freedom (d being the number of elements in the index set D), provided the index sets are selected to have no overlap in the components Y_x's. This test statistic can throw light on the nature of the local polynomial. If $\ell' = \ell + 1$, we can observe the progress of the sequence of F test statistics as ℓ is increased. When F dips below a critical level of F, we can tentatively identify the degree of the local polynomial. This degree can be built into the S measure in the Whittaker method and in the constraining polynomial in the MWA method. Because exposure to risk often changes slowly with age, a simple square or log transformation of the observations may stabilize the variance sufficiently.

Autocorrelation among the residuals after determining local, or perhaps global, polynomials may prevent attainment of the canonical assumptions. If the variance stabilizing transformation and the polynomial fitting produce, as would be expected, a stationary error series, the sample autocorrelation function (SACF) of the residuals can be calculated. The SACF can then be used to modify both the Whittaker and MWA methods.

In connection with the MWA method, a more general expression for the variance of the graduated terms is required. The expression (5.2) takes into consideration the stabilized variance and the autocorrelations.

$$\text{Var}\ (V_x) = \sigma^2 \left[\sum_{r=-m}^{m} a_r^2 + \sum_{r \neq s} \sum_{s} a_r a_s \rho_{|r-s|} \right] . \tag{5.2}$$

The expression within brackets would be minimized with the autocorrelations estimated from the SACF and the minimization done subject to the constraint of reproducing a polynomial of degree ℓ, where ℓ is estimated by the procedure described in connection with (5.1). The minimization is a standard quadratic programming problem.

Tables 1 and 2 illustrate these ideas. In Table 1 the error terms have an autoregression of order one form $(AR(1))$, i.e., $e_x = \phi e_{x-1} + a_x$, $|\ \phi\ | < 1$, and a_x have independent $N(0, \sigma^2)$ distributions. The autocorrelation of lag k is ϕ^k. In Table 2 the error term has a moving average of order one model (MA(1)), i.e.,

$$e_x = a_x - \theta a_{x-1}.$$

In Tables 1 and 2, 8-term MWA weights (coefficients) are displayed when $P_3(x)$ will be reproduced and the autocorrelations are as indicated. The symbol MVMWA denotes Minimum Variance Moving Weighted Average. Clearly adjusting for autocorrelations among error terms can make a difference.

From this viewpoint, minimum R_z formulas emerge naturally with a clear statistical motivation. If the error terms are of the form

$$e_x = (1 - B)^z a_x, \tag{5.3}$$

Table 1. *AR(1) Model*

	MWA Coefficients				Variances		
ϕ	a_0	a_1	a_2	a_3	MVMWA	Minimum$-R_0$	% Decrease
0.9	.254	.266	.246	−.139	$.9286\sigma^2$	$.9988\sigma^2$	7.03
0.8	.264	.260	.247	−.139	.8536	.9004	5.20
0.7	.271	.262	.237	−.134	.7778	.8085	3.80
0.6	.279	.265	.224	−.129	.7033	.7231	2.74
0.5	.288	.269	.210	−.123	.6314	.6439	1.94
0.4	.296	.274	.196	−.117	.5633	.5707	1.30
0.3	.307	.276	.182	−.112	.4993	.5033	0.79
0.2	.316	.280	.168	−.106	.4396	.4414	0.41
0.1	.325	.283	.155	−.100	.3844	.3848	0.10
0.0	.333	.286	.143	−.095	.3333	.3333	0.00
−0.1	.341	.289	.131	−.090	.2863	.2868	0.17
−0.2	.349	.292	.120	−.086	.2431	.2449	0.73
−0.3	.356	.294	.109	−.081	.2033	.2075	2.02
−0.4	.362	.297	.100	−.077	.1667	.1743	4.36
−0.5	.368	.299	.090	−.073	.1330	.1451	8.34
−0.6	.374	.301	.082	−.070	.1020	.1198	14.86
−0.7	.379	.303	.074	−.067	.0734	.0981	25.18
−0.8	.384	.305	.067	−.064	.0470	.0797	41.03
−0.9	.389	.307	.060	−.061	.0226	.0645	64.96

Table 2. *MA(1) Model*

ρ_1	θ	a_0	a_1	a_2	a_3	MVMWA	Minimum$-R_0$	% Decrease
0.5	−1.00	.476	.071	.357	−.166	$.5476\sigma^2$	$.5782\sigma^2$	5.29%
0.4	−0.50	.333	.231	.231	−.128	.5179	.5292	2.14
0.3	−0.33	.316	.264	.194	−.116	.4750	.4803	1.09
0.2	−0.21	.318	.276	.173	−.107	.4292	.4313	0.48
0.1	−0.10	.325	.282	.156	−.101	.3818	.3823	0.13
0.0	0.00	.333	.286	.143	−.095	.3333	.3333	0.00
−0.1	0.10	.341	.288	.132	−.091	.2840	.2844	0.13
−0.2	0.21	.347	.290	.122	−.086	.2338	.2354	0.68
−0.3	0.33	.357	.291	.113	−.083	.1831	.1864	1.78
−0.4	0.50	.364	.292	.105	−.079	.1318	.1374	4.08
−0.5	1.00	.372	.292	.097	−.076	.0801	.0885	9.49

where $Bf(x) = f(x-1)$ is the backward shift operator, the minimum variance MWA formula will be the minimum R_z formula. The selection of the minimum R_z formula will be guided by the fact that if model (5.3) is correct, the ACF will be given by

$$\rho_k = (-1)^k \frac{\binom{2z}{z-k}}{\binom{2z}{z}}, \qquad k = 0, 1, 2, \ldots, z.$$

Estimated autocorrelations can also be used to modify the Whittaker method. The expression for the F term is

$$(v-u)'W(v-u), \tag{5.4}$$

where W^{-1} is usually given a statistical interpretation as the variance-covariance matrix of U. If it appears that the autocorrelations are not all zero, the estimated values can be used in a respecification of W^{-1}.

In judging whether to include an a estimated autocorrelation in (5.2) or (5.4), we use the fact that if the error terms are uncorrelated, then we have the approximation $\mathrm{Var}(r_k) = 1/n$, where r_k is the estimate of the autocorrelation lag k and n is the number of terms in the series.

6. EXAMPLE

This example is taken from a thesis by Marks (1980). It is illustrative and does not constitute a complete analysis. The data is the 45 mortality probabilities shown by Miller (1946), page 65. Four preliminary variance stabilizing transformations are illustrated: (1) mortality ratio with the standard being the 1941 CSO table, (2) fourth root, (3) log, and (4) untransformed but with weighted least squares using the approximate weight $x^{-\frac{1}{3}}$.

Following the transformation of the data to stabilize the variance, a polynomial of minimum degree was determined for each set of transformed data. The SCAF for each set of residuals was calculated and in each case a moving average model of order 2 was indicated. The seventeen term symmetric MWA formulas for minimizing the variance of the graduated terms with the covariance structure indicated by the SACF was determined by quadratic programming. The resulting coefficients are shown in Table 3.

Results of the graduations are presented in Table 4 and certain standard tests of graduation in Table 5. By these measures the mortality ratio transformation with the adjustment to recognize the covariance structure is the champion.

Table 3. *MWA Coefficient for Example*

	Mortality Ratio	Fourth Root	Log	Untransformed	Minimum$-R_0$
a_0	.22563849	.22547874	.22484624	.21167833	.21028808
a_1	.20708740	.20697204	.20636317	.19509355	.19647542
a_2	.15666693	.15667748	.15667733	.15185186	.15718034
a_3	.08636105	.08647163	.08672577	.11251874	.09983309
a_4	.01637196	.01655627	.01775453	.02828609	.03215051
a_5	−.03605064	−.03601728	−.03611529	−.03623343	−.02786380
a_6	−.04730217	−.04729007	−.04600439	−.06128083	−.06191954
a_7	−.03148169	−.03180314	−.03480732	−.03953373	−.04643966
a_8	.03552799	.03569379	.03698318	.04345852	.04643967
V^*	.0513	.0527	.0572	.0635	.2103

Note: V^* is min $\mathrm{Var}(V_x)/\sigma^2$ for the covariance structure assumed.

Table 4. *Graduated Values (Deaths per Thousand)*

Age	Ungraduated Values	Graduated Values: Mortality Ratio	Fourth Root	Log	Untrans-formed	Minimum R_0
43	3.11	3.82	3.82	3.82	3.83	3.83
44	3.65	4.53	4.53	4.54	4.60	4.64
45	4.46	5.17	5.17	5.17	5.26	5.22
46	6.32	6.05	6.05	6.05	6.09	6.06
47	7.41	6.57	6.57	6.56	6.46	6.48
48	7.26	7.53	7.53	7.55	7.44	7.49
49	9.45	8.00	8.00	7.97	8.02	7.97
50	7.49	8.41	8.41	8.42	8.40	8.37
51	7.63	8.79	8.79	8.80	8.82	8.83
52	10.64	9.56	9.56	9.57	9.59	9.66
53	9.99	11.19	11.20	11.22	11.41	11.40
54	13.78	12.07	12.07	12.04	12.04	11.98
55	9.67	12.85	12.85	12.84	12.54	12.48
56	18.26	15.06	15.07	15.10	15.05	15.21
57	18.11	16.14	16.13	16.10	16.22	16.20
58	15.93	18.16	18.17	18.21	18.29	18.20
59	17.89	20.04	20.04	20.02	20.15	20.16
60	18.53	21.88	21.88	21.88	21.70	21.75
61	32.46	24.69	24.70	24.71	24.91	24.83
62	27.94	26.92	26.92	26.91	26.77	26.93
63	19.37	28.77	28.77	28.76	28.54	28.47
64	40.00	29.94	29.93	29.88	29.86	29.58
65	27.95	32.87	32.87	32.94	32.87	33.13
66	37.64	36.12	36.12	36.12	36.44	36.69
67	41.23	40.21	40.21	40.23	40.61	40.31
68	34.59	44.94	44.94	44.93	44.98	44.85
69	53.91	50.56	50.56	50.57	50.53	50.88
70	57.21	51.00	50.99	50.90	50.32	49.94
71	65.04	54.06	54.06	54.10	53.31	53.48

7. SUMMARY

MWA and Whittaker graduation methods have both mathematical programming and statistical foundations. If the statistical foundations are taken seriously, techniques from applied statistics to bring the analysis into accord

Table 5. *Characteristics of Sample Graduations*

Characteristic Statistics	Mortality Ratio	Fourth Root	Log	Untrans-formed	Minimum R_0
$\sum_x (v_x - u_x)^2$	613.61	613.86	615.37	632.12	635.08
$\sum_x (\Delta v_x)^2$	142.20	142.32	143.49	142.59	152.80
$\sum_x (\Delta^2 v_x)^2$	47.15	47.66	51.72	63.48	93.65
$\sum_x (\Delta^3 v_x)^2$	129.94	131.56	145.03	181.65	266.38
$\sum_x (\Delta^4 v_x)^2$	338.84	343.37	383.79	490.02	681.59
$\sum_x (\Delta^5 v_x)^2$	823.07	835.66	960.12	1255.10	1601.40

with the model may be necessary. These techniques include variance stabilizing transformations and incorporating correlated error terms into variance estimates. Iteration of the steps may be necessary to achieve acceptable results.

REFERENCES

Abrahams, B., and H. Ledolter (1983), *Statistical Methods for Forecasting.* New York: Wiley and Sons.

Andrews, G., and C. J. Nesbitt (1965), "Periodograms of graduation operators." *Transactions of the Society of Actuaries* **17**, 166–177.

Borgan, O. (1979), "On the theory of moving average graduation." *Scandinavian Actuarial Journal*, 83–105.

Chan, F., L. Chan, and E. Mead (1982), "Properties and modifications of Whittaker-Henderson graduation." *Scandinavian Actuarial Journal*, 368–369.

Chan, F., L. Chan, and M. Yu (1985), "A generalization of Whittaker-Henderson graduation." *Transactions of the Society of Actuaries* **36**.

Chan, L., and H. Panjer (1983), "A statistical approach to graduation by mathematical formula." *Insurance Mathematics and Economics* **2**, 33–48.

Craven, P., and G. Wahba (1979), "Smoothing noisy data with spline functions." *Numeriske Mathematik* **31**, 377–403.

Greville, T. (1973), *Part 5 Study Notes: Graduation.* Chicago: Society of Actuaries.

Henderson, R. (1938), *Mathematical Theory of Graduation.* New York: Actuarial Society of America.

Hickman, J. (1967), "Discussion of Bayesian graduation." *Transactions of the Society of Actuaries* **19,** 114–116.

Hickman, J., and R. Miller (1977), "Notes on Bayesian graduation." *Transactions of the Society of Actuaries* **29,** 7–21.

Hoem, J. (1984), "A contribution to the theory of linear graduation." *Insurance: Mathematics and Economics* **3,** 66–112.

Kimeldorf, G., and D. Jones (1967), "Bayesian graduation." *Transactions of the Society of Actuaries* **19,** 66–112.

London, R. (1981), "In defense of minimum R linear compound graduation, and a simple modification for its improvement." *Actuarial Research Clearing House* 1981.2, 75–88.

London, R. (1984), *Graduation: The Revision of Estimates.* Itasca, Illinois: Society of Actuaries.

Marks, S. (1980), "A statistical view of actuarial graduation incorporating a correlated error structure." Ph.D. Thesis, University of Wisconsin-Madison.

Miller, M. (1946), *Elements of Graduation.* New York and Chicago: Actuarial Society of American and American Institute of Actuaries.

Schoenberg, I. (1964), "Spline functions and the problem of graduation." *Proceedings of the National Academy of Science* **52,** 947–950.

Schuette, D. (1978), "A linear programming approach to graduation." *Transactions of the Society of Actuaries* **30,** 407–431.

Sheppard, W. (1914), "Graduation by reduction of mean square error." *Journal of the Institute of Actuaries* **48,** 117–185.

Shiu, E. (1985), "Minimum R_z moving weighted average formulas." *Transactions of the Society of Actuaries* **36.**

Tenenbein, A., and I. Vanderhoof (1980), "New mathematical laws of select and ultimate mortality." *Transactions of the Society of Actuaries* **32,** 119–158.

Whittaker, E. (1923), "A new method of graduation." *Proceedings of the Edinburgh Mathematical Society* **41,** 63–75.

Whittaker, E., and G. Robinson (1924), *The Calculus of Observations.* London: Blackie and Sons.

James D. Broffitt [1]

ISOTONIC BAYESIAN GRADUATION WITH AN ADDITIVE PRIOR

ABSTRACT

Assume the mortality rate at age $x+j-1$ is $q_{x+j-1} = 1 - \exp(-\theta_j)$, $j = 1,\ldots,k$. Isotonic Bayesian graduation provides a Bayes estimator of $\theta_1,\ldots,\theta_k$ (and consequently, $q_x,\ldots,q_{x+k-1}$) under the assumption $\theta_1 < \cdots < \theta_k$. This is accomplished by specifying a prior distribution for which $P(\Theta_1 < \cdots < \Theta_k) = 1$. In a previous paper the prior was defined by $(\Theta_1,\ldots,\Theta_k) \stackrel{D}{=} (Y_1,\cdots,Y_k \mid Y_1 < \cdots < Y_k)$ where $Y_1,\ldots,Y_k$ are independent. In this paper the Bayes estimator is developed using the prior $(\Theta_1,\ldots,\Theta_k) \stackrel{D}{=} (Y_1, Y_1+Y_2,\ldots,Y_1+\cdots+Y_k)$. The advantages are an easier specification of the prior parameters and shorter computational time.

1. INTRODUCTION

The graduation of mortality rates has been traditionally viewed as a technique for computing a modified series of estimates which possess a high degree of smoothness while maintaining reasonable fit to the initial series of estimates. These techniques have usually been developed from a data oriented viewpoint without regard for a statistical model of the data. Graduation techniques have been compared by computing measures of smoothness, and fit of the graduated values to the initial estimates, without considering how these techiques would compare in repeated samples. Lacking the influence of a model, the graduated values may be too closely tied to the observed data. In the event of an unrepresentative sample, these graduations can produce quite imprecise estimates of the true rates. (This possibility is less likely for graduation techniques which utilize prior information.)

[1] Department of Statistics and Actuarial Science, The University of Iowa, Iowa City, Iowa 52242

I. B. MacNeill and G. J. Umphrey (eds.), Actuarial Science, 19–39.

The selection of an initial estimator and a graduation technique should be treated as a single activity, namely, the process of solving a problem in statistical estimation. Let $q_x, \ldots, q_{x+k-1}$ be the mortality rates at ages $x, \ldots, x+k-1$ for a particular population. Then $q_x, \ldots, q_{x+k-1}$ is a set of parameters which have specific but unknown values which we desire to estimate. A definitive analysis demands that a statistical model be specified for the data which incorporates an adequate amount of prior knowledge about these parameters. If the data is analyzed in accord with this model, the resulting estimates (graduated values) should not need modification. The secondary process of graduation will automatically occur. Moreover, these estimates will be influenced by both the data and the model. Inaccuracies caused by an unrepresentative sample should be less severe than for graduations that are based totally on the observed data.

It is necessary to specify a model which incorporates the notions of smoothness, but is not so complex that analysis is impractical. In this paper smoothness is achieved by assuming the parameters are increasing, i.e., $q_x < \cdots < q_{x+k-1}$. This is known as the isotonic assumption.

Let $\mu(y)$ be the force of mortality at age y. We assume $\mu(y)$ is constant over one-year periods. Thus for $j = 1, \ldots, k$,

$$\mu(y) = \theta_j, \qquad x+j-1 \le y < x+j, \tag{1.1}$$

and consequently,

$$q_{x+j-1} = 1 - \exp(-\theta_j). \tag{1.2}$$

Then the isotonic restriction $q_x < \cdots < q_{x+k-1}$ is equivalent to $\theta_1 < \cdots < \theta_k$. Our procedure is to first estimate θ_j, and then q_{x+j-1} by substitution into (1.2).

We shall develop Bayes estimators for $\theta_1, \ldots, \theta_k$ with respect to prior distributions for which $P(\Theta_1 < \cdots < \Theta_k) = 1$. This guarantees that $P(\Theta_1 < \cdots < \Theta_k|\text{Data}) = 1$, i.e., the forces of mortality are increasing with probability one in the posterior distribution. Assuming squared error loss the isotonic Bayes estimator of θ_j is $\theta_j^{IB} = E(\Theta_j|\text{Data})$. Thus $\theta_1^{IB} < \cdots < \theta_k^{IB}$, and accordingly $q_x^{IB} < \cdots < q_{x+k-1}^{IB}$ where $q_{x+j-1}^{IB} = 1 - \exp(-\theta_j^{IB})$ denotes the corresponding estimator of q_{x+j-1}.

In an earlier paper (Broffitt, 1984b), the isotonic Bayes estimators were developed for a prior distribution of the form

$$(\Theta_1, \ldots, \Theta_k) \stackrel{D}{=} (Y_1, \ldots, Y_k | Y_1 < \cdots < Y_k),$$

where $Y_1, \ldots, Y_k$ are independent gamma random variables. We shall refer to this as the *ordered prior*. Specifying the parameters of these gamma distributions posed certain difficulties which required time consuming trial and error calculations.

In this paper the isotonic Bayes estimators will be developed using priors defined by

$$(\Theta_1, \ldots, \Theta_k) \stackrel{D}{=} (Y_1, Y_1 + Y_2, \ldots, Y_1 + \cdots + Y_k),$$

where $Y_1, \ldots, Y_k$ are independent gamma random variables. This will be called the *additive prior*. The advantages of the additive prior are an easier specification of the prior parameters and, in a special case, shorter computational time.

2. THE LIKELIHOOD FUNCTION

We are interested in mortality rates between ages x and $x+k$. Suppose N lives come under observation at some point within this age interval. Let $x+a_i$ be the age where life i begins observation, and $x+t_i$ be the age where observation ceases. This cessation of observation may be caused by death, voluntary withdrawal, the attainment of age $x+k$, or termination of the observation period. Under a model that assumes independent random times to death and withdrawal, the likelihood function is

$$L \propto \prod_{i \in D} \mu(x+t_i) \cdot \exp\left[-\sum_{i=1}^{N} \int_{x+a_i}^{x+t_i} \mu(y)dy\right], \tag{2.1}$$

where D denotes the subset of subscripts corresponding to those lives that died. (For further elaboration see Broffitt (1984a,b), Steelman (1968), and Chan and Panjer (1983).)

We now utilize assumption (1.1) to simplify (2.1). Letting d_j equal the number of deaths observed between ages $x+j-1$ and $x+j$,

$$\prod_{i \in D} \mu(x+t_i) = \prod_{j=1}^{k} \theta_j^{d_j}.$$

Also let e_{ij} equal the amount of time, measured in years, that life i was under observation between ages $x+j-1$ and $x+j$. Then

$$\begin{aligned}\sum_{i=1}^{N}\left[\int_{x+a_i}^{x+t_i} \mu(y)dy\right] &= \sum_{i=1}^{N}\sum_{j=1}^{k} e_{ij}\theta_j \\ &= \sum_{j=1}^{k} e_j\theta_j,\end{aligned}$$

where $e_j = \sum_{i=1}^{N} e_{ij}$ is the total number of years, the N lives were under observation between ages $x+j-1$ and $x+j$. Thus (2.1) simplifies to

$$L(\theta_1,\ldots,\theta_k) \propto \prod_{j=1}^{k}\left[\theta_j^{d_j} e^{-e_j\theta_j}\right]. \tag{2.2}$$

The unrestricted MLE of θ_j is $\theta_j^M = d_j/e_j$, which corresponds to the well known result in reliability theory that when lifetime is exponentially distributed, the hazard rate is estimated by total failures divided by total time on test. In our problem we shall refer to e_j as "exposure".

Finally we note that the isotonic Bayes estimators, to be developed, are not particular to the mortality rate problem, but apply whenever the likelihood has the same form as (2.2). For example, suppose $\theta_1,\ldots,\theta_k$ are the parameters for k independent Poisson distributions. If a sample of size e_j is taken from the jth distribution and the sample total is d_j, then the likelihood is given by (2.2).

3. RESULTS FOR THE ORDERED PRIOR

In this section we shall summarize some items given by Broffitt (1984b), but first some notation. The prior and posterior pdf's will be denoted by $prior_O(\theta_1,\ldots,\theta_k)$ and $post_O(\theta_1,\ldots,\theta_k)$, respectively, when using the ordered prior. The subscript "A" will be used when referring to the additive prior. Correspondingly, θ^{IBO} and θ^{IBA} will denote isotonic Bayes estimators with respect to the ordered and additive priors respectively. The symbols $X \sim G(\alpha,\beta)$ mean that X has the gamma pdf

$$g(x;\alpha,\beta) = \begin{cases} \beta^\alpha x^{\alpha-1}e^{-\beta x}/\Gamma(\alpha), & x>0 \\ 0, & \text{otherwise.} \end{cases}$$

Now let $Y_1,\ldots,Y_k$ be independent where $Y_j \sim G(\delta_j,\varepsilon_j)$. The ordered prior is defined by

$$(\Theta_1,\ldots,\Theta_k) \stackrel{D}{=} (Y_1,\ldots,Y_k | Y_1 < \cdots < Y_k).$$

Consequently,

$$prior_O(\theta_1,\ldots,\theta_k) \propto \prod_{j=1}^{k} g(\theta_j;\delta_j,\varepsilon_j), \quad \theta_1 < \cdots < \theta_k,$$

and

$$post_0(\theta_1,\ldots,\theta_k) \propto L(\theta_1,\ldots,\theta_k)\; prior_0(\theta_1,\ldots,\theta_k)$$
$$\propto \prod_{j=1}^{k} g(\theta_j;\alpha_j,\lambda_j), \quad \theta_1 < \cdots < \theta_k, \tag{3.1}$$

where $\alpha_j = \delta_j + d_j$ and $\lambda_j = \varepsilon_j + e_j$. Thus

$$(\Theta_1,\ldots,\Theta_k|\text{Data}) \stackrel{D}{=} (X_1,\ldots,X_k|X_1 < \cdots < X_k),$$

where $X_1,\ldots,X_k$ are independent and $X_j \sim G(\alpha_j,\lambda_j)$.

Let $\boldsymbol{\alpha} = (\alpha_1,\ldots,\alpha_k)$, $\boldsymbol{\lambda} = (\lambda_1,\ldots,\lambda_k)$, $\boldsymbol{\alpha}^{(j)} = (\alpha_1,\ldots,\alpha_{j-1},\alpha_j + 1,\alpha_{j+1},\ldots,\alpha_k)$, and $R(\boldsymbol{\alpha};\boldsymbol{\lambda}) = P(X_1 < \cdots < X_k)$. Then

$$post_0(\theta_1,\ldots,\theta_k) = \frac{\prod_{j=1}^{k} g(\theta_j;\alpha_j,\lambda_j)}{R(\boldsymbol{\alpha};\boldsymbol{\lambda})}, \qquad \theta_1 < \cdots < \theta_k,$$

Finally, θ_j^{IBO}, obtained by integrating $\theta_j\; post_0(\theta_1,\ldots,\theta_k)$, is given in the following theorem.

Theorem 1. $\theta_j^{IBO} = (\alpha_j/\lambda_j)R(\boldsymbol{\alpha}^{(j)};\boldsymbol{\lambda})/R(\boldsymbol{\alpha};\boldsymbol{\lambda})$.

Before computing this estimate it is necessary to specify $\delta_1,\ldots,\delta_k,\varepsilon_1,\ldots,\varepsilon_k$. Let $\theta_1^0,\ldots,\theta_k^0$ be prior guesses for the values of $\theta_1,\ldots,\theta_k$. Then one set of conditions we would like to satisfy is

$$\theta_j^0 = E(\Theta_j)$$
$$= \frac{\delta_j}{\varepsilon_j}\frac{R(\boldsymbol{\delta}^{(j)};\boldsymbol{\varepsilon})}{R(\boldsymbol{\delta};\boldsymbol{\varepsilon})}, \qquad j = 1,\ldots,k,$$

where $\boldsymbol{\delta} = (\delta_1,\ldots,\delta_k)$, $\boldsymbol{\varepsilon} = (\varepsilon_1,\ldots,\varepsilon_k)$ and $\boldsymbol{\delta}^{(j)} = (\delta_1,\ldots,\delta_{j-1},\delta_j + 1,\delta_{j+1},\ldots,\delta_k)$. This requires selecting the prior parameters by trial and error. For each set of parameters tried, $R(\boldsymbol{\delta};\boldsymbol{\varepsilon})$, $R(\boldsymbol{\delta}^{(1)};\boldsymbol{\varepsilon}),\ldots,R(\boldsymbol{\delta}^{(k)};\boldsymbol{\varepsilon})$ must be computed. This computation can become quite time consuming, especially for sizeable values of k. The algorithm used for computing $R(\cdot;\cdot)$ is based on the following representation which is restricted to the case where $\alpha_2,\ldots,\alpha_k$ are integers:

$$R(\boldsymbol{\alpha};\boldsymbol{\lambda}) = \sum_{i_{k-1}=0}^{\alpha_k-1} f(i_{k-1};\alpha_{k-1},\rho_{k-1}) \sum_{i_{k-2}=0}^{i_{k-1}+\alpha_{k-1}-1} f(i_{k-2};\alpha_{k-2},\rho_{k-2})$$
$$\sum_{i_1=0}^{i_2+\alpha_2-1} f(i_1;\alpha_1,\rho_1),$$

where $\rho_j = \lambda_j/(\lambda_j + \cdots + \lambda_k)$ and $f(i;\alpha,\rho) = \frac{\Gamma(i+\alpha)}{i!\Gamma(\alpha)}\rho^\alpha(1-\rho)^i$.

4. RESULTS FOR THE ADDITIVE PRIOR

Consider the parameterization defined by the following equations:

$$\begin{array}{ll} \phi_1 = \theta_1 & \theta_1 = \phi_1 \\ \phi_2 = \theta_2 - \theta_1 & \theta_2 = \phi_1 + \phi_2 \\ \vdots & \vdots \\ \phi_k = \theta_k - \theta_{k-1} & \theta_k = \phi_1 + \phi_2 + \cdots + \phi_k. \end{array} \tag{4.1}$$

The parameters $\phi_2, \ldots, \phi_k$ represent increments in the θ's. This transformation is one-to-one between the sets

$$\{(\theta_1, \ldots, \theta_k); \qquad 0 < \theta_1 < \cdots < \theta_k < \infty\}$$

and

$$\{(\phi_1, \ldots, \phi_k); \qquad \phi_1 > 0, \ldots, \phi_k > 0\},$$

and the Jacobian is one. Thus we can work with either set of parameters, and may easily switch from one to the other by simply substituting according to equations (4.1).

Let $Y_1, \ldots, Y_k$ be independent where $Y_j \sim G(a_j, r_j)$. The additive prior distribution defined by

$$(\Theta_1, \ldots, \Theta_k) \stackrel{D}{=} (Y_1, Y_1 + Y_2, \ldots Y_1 + Y_2 + \cdots + Y_k) \tag{4.2}$$

is clearly equivalent to the prior on $\Phi_1, \ldots, \Phi_k$ given by

$$(\Phi_1, \ldots, \Phi_k) \stackrel{D}{=} (Y_1, \ldots, Y_k).$$

Thus

$$prior_A(\phi_1, \ldots, \phi_k) \propto \prod_{j=1}^{k} \phi_j^{a_j - 1} e^{-r_j \phi_j}, \quad \phi_j > 0, \quad j = 1, \ldots, k$$

whence

$$\begin{aligned} &post_A(\phi_1, \ldots, \phi_k) \\ &\quad \propto \prod_{j=1}^{k} (\phi_1 + \cdots + \phi_j)^{d_j} e^{-e_j(\phi_1 + \cdots + \phi_j)} prior_A(\phi_1, \ldots, \phi_k) \\ &\quad \propto \prod_{j=1}^{k} (\phi_1 + \cdots + \phi_j)^{d_j} \phi_j^{a_j - 1} e^{-b_j \phi_j}, \quad \phi_j > 0, \quad j = 1, \ldots, k, \end{aligned}$$

where $b_j = r_j + e_j + \cdots + e_k$. Therefore

$$post_A(\phi_1, \ldots, \phi_k) \propto \prod_{j=1}^{k} (\phi_1 + \cdots + \phi_j)^{d_j} g(\phi_j; a_j, b_j). \tag{4.3}$$

Let $Z_1, \ldots, Z_k$ be independent where $Z_j \sim G(a_j, b_j)$, $\mathbf{d} = (d_1, \ldots, d_k)$, $\mathbf{a} = (a_1, \ldots, a_k)$, $\mathbf{b} = (b_1, \ldots, b_k)$ and define

$$B(\mathbf{d}; \mathbf{a}; \mathbf{b}) = E[Z_1^{d_1}(Z_1 + Z_2)^{d_2} \ldots (Z_1 + \cdots + Z_k)^{d_k}]. \tag{4.4}$$

It follows from (4.3) that

$$post_A(\phi_1, \ldots, \phi_k) = \frac{\prod_{j=1}^{k} (\phi_1 + \cdots + \phi_j)^{d_j} g(\phi_j; a_j, b_j)}{B(\mathbf{d}; \mathbf{a}; \mathbf{b})}.$$

Now $E(\Theta_j|\text{Data}) = \sum_{i=1}^{j} E(\Phi_i|\text{Data})$, that is,

$$\theta_j^{IBA} = \sum_{i=1}^{j} \phi_i^{IBA},$$

where

$$\begin{aligned} \phi_i^{IBA} &= E(\Phi_i|\text{Data}) \\ &= \int_0^\infty \cdots \int_0^\infty \phi_i post_A(\phi_1, \ldots, \phi_k) d\phi_1 \ldots d\phi_k \\ &= \frac{a_i}{b_i} \cdot \frac{B(\mathbf{d}; \mathbf{a}^{(i)}; \mathbf{b})}{B(\mathbf{d}; \mathbf{a}; \mathbf{b})}, \end{aligned} \tag{4.5}$$

where $\mathbf{a}^{(i)} = (a_1, \ldots, a_{i-1}, a_i + 1, a_{i+1}, \ldots, a_k)$. Equation (4.5) follows since

$$\phi g(\phi; a, b) = \frac{a}{b} g(\phi; a + 1, b).$$

This development is summarized in Theorem 2.

Theorem 2. For the likelihood (2.2) the Bayes estimator of θ_j with respect to the additive prior (4.2) is

$$\theta_j^{IBA} = \sum_{i=1}^{j} \frac{a_i}{b_i} \frac{B(\mathbf{d}; \mathbf{a}^{(i)}; \mathbf{b})}{B(\mathbf{d}; \mathbf{a}; \mathbf{b})},$$

where $B(\cdot;\cdot;\cdot)$ is defined by (4.4).

Notice that the prior parameters $a_1, \ldots, a_k, r_1, \ldots, r_k$ may be easily determined once the prior mean and variance of Θ_j, $j = 1, \ldots, k$ have been selected. Let θ_j^0 and v_j^0 be the desired values of the prior mean and variance of Θ_j, respectively. Notice that for the additive prior we must have $v_1^0 < \cdots < v_k^0$. This does not seem to be a severe restriction since it is natural for the variability of Θ_j to increase as its expectation increases. Also for the older ages our prior information is probably less certain. Then

$$\begin{aligned} \theta_j^0 - \theta_{j-1}^0 &= E(\Phi_j) \\ & a_j/r_j, \quad j = 1, \ldots, k, \end{aligned}$$

and

$$\begin{aligned} v_j^0 - v_{j-1}^0 &= \mathrm{Var}(\Phi_j) \\ &= a_j/r_j^2, \quad j = 1, \ldots, k, \end{aligned}$$

where $\theta_0^0 = v_0^0 = 0$. Solving these equations yields

$$r_j = \frac{\theta_j^0 - \theta_{j-1}^0}{v_j^0 - v_{j-1}^0},$$

and

$$a_j = \frac{(\theta_j^0 - \theta_{j-1}^0)^2}{v_j^0 - v_{j-1}^0}.$$

The only difficulty in applying Theorem 2 is in the calculation of $B(\mathbf{d}; \mathbf{a}; \mathbf{b})$. It may be possible to develop an efficient algorithm or a good approximation for calculating $B(\cdot;\cdot;\cdot)$, however, there is a special case for which θ_j^{IBA} may be computed using an algorithm analogous to that employed in computing θ_j^{IBO}.

4.1 A Special Case

Let $\lambda_j = e_j + r_j - r_{j+1}$, $j = 1, \ldots, k$ where $r_{k+1} = 0$, and notice that $\sum_{j=1}^k b_j \phi_j = \sum_{j=1}^k \lambda_j \theta_j$. Using this together with (4.3) and (4.1), we obtain the posterior pdf of $\Theta_1, \ldots, \Theta_k$ as

$$post_A(\theta_1, \ldots, \theta_k) \propto \prod_{j=1}^{k} \theta_j^{d_j} (\theta_j - \theta_{j-1})^{a_j - 1} e^{-\lambda_j \theta_j}, \quad 0 < \theta_1 < \cdots < \theta_k \tag{4.6}$$

where $\theta_0 = 0$. Now assume $a_2 = \cdots = a_k = 1$ so that $Y_2, \ldots, Y_k$ have exponential distributions, and let $\alpha_j = a_j + d_j$. Then (4.6) reduces to

$$post_A(\theta_1, \ldots, \theta_k) \propto \prod_{j=1}^{k} \theta_j^{\alpha_j - 1} e^{-\lambda_j \theta_j}, \quad 0 < \theta_1 < \cdots < \theta_k. \tag{4.7}$$

On comparing (4.7) with (3.1) it is tempting to conclude that when $a_2 = \cdots = a_k = 1$, the posterior distribution of $\Theta_1, \ldots, \Theta_k$ with respect to the additive prior and the ordered prior have the same form. That is

$$(\Theta_1, \ldots, \Theta_k | \text{Data}) \overset{D}{=} (X_1, \ldots, X_k | X_1 < \cdots < X_k), \tag{4.8}$$

where $X_1, \ldots, X_k$ are independent and $X_j \sim G(\alpha_j, \lambda_j)$. This is precisely the case *if* $\lambda_j > 0$, $j = 1, \ldots, k$. This case does occur but it is also quite possible for λ_j to be negative for one or more values of j. When this happens (4.7) is still valid but (4.8) does not follow.

From (4.7) we obtain

$$post_A(\theta_1, \ldots, \theta_k) \propto \prod_{j=1}^{k} |\lambda_j|^{\alpha_j} \theta_j^{\alpha_j - 1} e^{-\lambda_j \theta_j} / \Gamma(\alpha_j), \quad 0 < \theta_1 < \cdots < \theta_k.$$

Define

$$g^+(x; \alpha, \lambda) = \begin{cases} |\lambda|^{\alpha} x^{\alpha-1} e^{-\lambda x} / \Gamma(\alpha), & x > 0 \\ 0, & \text{otherwise,} \end{cases}$$

and let

$$R^+(\boldsymbol{\alpha}; \boldsymbol{\lambda}) = \int_0^\infty \int_{x_1}^\infty \cdots \int_{x_{k-1}}^\infty \left[\prod_{j=1}^{k} g^+(x_j; \alpha_j, \lambda_j) \right] dx_k \cdots dx_2 dx_1. \tag{4.9}$$

Then

$$post_A(\theta_1, \ldots, \theta_k) = \prod_{j=1}^{k} \frac{g^+(\theta_j; \alpha_j, \lambda_j)}{R^+(\boldsymbol{\alpha}; \boldsymbol{\lambda})}, \quad \theta_1 < \cdots < \theta_k,$$

from which the estimator $E(\Theta_j | \text{Data})$ is easily obtained by integration.

Theorem 3. Under the additive prior with $a_2 = \cdots = a_k = 1$,

$$\theta_j^{IBA} = \frac{\alpha_j}{|\lambda_j|} \frac{R^+(\boldsymbol{\alpha}^{(j)}; \boldsymbol{\lambda})}{R^+(\boldsymbol{\alpha}; \boldsymbol{\lambda})},$$

where $R^+(\cdot; \cdot)$ is defined in (4.9).

4.2 Computation of $R^+(\boldsymbol{\alpha}; \boldsymbol{\lambda})$

In order to compute θ_j^{IBA} using Thereom 3 it is necessary to calculate $R^+(\boldsymbol{\alpha}; \boldsymbol{\lambda})$. Corollary 4.1 below provides a formula for this purpose.

For notational convenience let

$$R_k^+(w;\boldsymbol{\alpha};\boldsymbol{\lambda}) = \int_w^\infty \int_{x_1}^\infty \cdots \int_{x_{k-1}}^\infty \left[\prod_{j=1}^k g^+(x_j;\alpha_j,\lambda_j)\right] dx_k \ldots dx_2 dx_1,$$

and notice that $R^+(\boldsymbol{\alpha};\boldsymbol{\lambda}) = R_k^+(0,\boldsymbol{\alpha};\boldsymbol{\lambda})$.

Let

$$\overline{G}(w;\alpha,\lambda) = \int_w^\infty g(x;\alpha,\lambda)dx.$$

If α is an integer, then it is well known that

$$\overline{G}(x;\alpha,\lambda) = \sum_{i=0}^{\alpha-1} (\lambda x)^i e^{-\lambda x}/i!. \tag{4.10}$$

Also let $\rho_j = \lambda_j/(\lambda_j + \cdots + \lambda_k)$, $j = 1,\ldots,k-1$, and let

$$f^+(i;\alpha,\rho) = \frac{\Gamma(i+\alpha)}{i!\Gamma(\alpha)}|\rho|^\alpha(1-\rho)^i.$$

Lemma 1. If $k=2$, $\lambda_1+\lambda_2 > 0$, $\lambda_2 > 0$ and α_2 is an integer, then

$$R_2^+(w;\boldsymbol{\alpha};\boldsymbol{\lambda}) = \sum_{i=0}^{\alpha_2-1} f^+(i;\alpha_1,\rho_1)\overline{G}(w;i+\alpha_1,\lambda_1+\lambda_2).$$

Proof.

$$\begin{aligned} R_2^+(w;\boldsymbol{\alpha};\boldsymbol{\lambda}) &= \int_w^\infty g^+(x_1;\alpha_1,\lambda_1)\overline{G}(x_1;\alpha_2,\lambda_2)dx_1 \\ &= \sum_{i=0}^{\alpha_2-1} \int_w^\infty f^+(i;\alpha_1,\rho_1)g(x_1;i+\alpha_1,\lambda_1+\lambda_2)dx_1 \end{aligned}$$

by (4.10), *QED*.

Theorem 4. For $k \geq 2$, if $\lambda_j + \cdots + \lambda_k > 0$, $j = 1,\ldots,k$ and $\alpha_2,\ldots,\alpha_k$ are integers, then

$$\begin{aligned} &R_k^+(w;\boldsymbol{\alpha};\boldsymbol{\lambda}) \\ &= \sum_{i_{k-1}=0}^{\alpha_k-1} f^+(i_{k-1};\alpha_{k-1},\rho_{k-1}) \sum_{i_{k-2}=0}^{i_{k-1}+\alpha_{k-1}-1} f^+(i_{k-2};\alpha_{k-2},\rho_{k-2}) \\ &\quad \cdot \sum_{i_1=0}^{i_2+\alpha_2-1} f^+(i_1;\alpha_1,\rho_1)\overline{G}(w;i_1+\alpha_1,\lambda_1+\cdots+\lambda_k). \end{aligned} \tag{4.11}$$

Proof. The proof is by induction on k. For $k = 2$ the theorem is clearly true since it is the same as Lemma 1. Now assume the theorem is true for a general k and prove it holds for $k+1$. By definition

$$\begin{aligned} R^+_{k+1}(w;\boldsymbol{\alpha};\boldsymbol{\lambda}) &= \int_w^\infty \int_{x_1}^\infty \cdots \int_{x_k}^\infty \left[\prod_{j=1}^{k+1} g^+(x_j;\alpha_j,\lambda_j)\right] dx_{k+1}\ldots dx_2 dx_1 \\ &= \int_w^\infty g^+(x_1;\alpha_1,\lambda_1) \\ &\cdot \left\{\int_{x_1}^\infty \cdots \int_{x_k}^\infty \left[\prod_{j=2}^{k+1} g^+(x_j;\alpha_j,\lambda_j)\right] dx_{k+1}\ldots dx_2\right\} dx_1. \end{aligned} \tag{4.12}$$

The term in braces in (4.12) is $R^+_k(x_1;\boldsymbol{\alpha};\boldsymbol{\lambda})$, where $\boldsymbol{\alpha} = (\alpha_2,\ldots,\alpha_{k+1})$ and $\boldsymbol{\lambda} = (\lambda_2,\ldots,\lambda_{k+1})$. Using (4.11),

$$\begin{aligned} R^+_k(x_1;\boldsymbol{\alpha};\boldsymbol{\lambda}) &= \sum_{i_k=0}^{\alpha_{k+1}-1} f^+(i_k;\alpha_k,\rho_k) \sum_{i_{k-1}=0}^{i_k+\alpha_k-1} f^+(i_{k-1};\alpha_{k-1},\rho_{k-1}) \\ &\cdot \sum_{i_2=0}^{i_3+\alpha_3-1} f^+(i_2;\alpha_2,\rho_2)\overline{G}(x_1;i_2+\alpha_2,\lambda_2+\cdots+\lambda_{k+1}), \end{aligned}$$

where $\rho_j = \lambda_j/(\lambda_j+\cdots+\lambda_{k+1})$ in this case. The proof is completed by substituting this expression into (4.12), and noting that as in Lemma 1,

$$\begin{aligned} &\int_w^\infty g^+(x_1;\alpha_1,\lambda_1)\overline{G}(x_1;i_2+\alpha_2,\lambda_2+\cdots+\lambda_{k+1})dx_1 \\ &\qquad = \sum_{i_1=0}^{i_2+\alpha_2-1} f^+(i_1;\alpha_1,\rho_1)\overline{G}(w;i_1+\alpha_1,\lambda_1+\cdots+\lambda_{k+1}). \end{aligned}$$

Corollary 4.1. Under the conditions of Theorem 4,

$$\begin{aligned} R^+(\boldsymbol{\alpha};\boldsymbol{\lambda}) &= \sum_{i_{k-1}=0}^{\alpha_k-1} f^+(i_{k-1};\alpha_{k-1},\rho_{k-1}) \sum_{i_{k-2}=0}^{i_{k-1}+\alpha_{k-1}-1} f^+(i_{k-2};\alpha_{k-2},\rho_{k-2}) \\ &\cdot \sum_{i_1=0}^{i_2+\alpha_2-1} f^+(i_1;\alpha_1,\rho_1). \end{aligned}$$

Proof. Let $w = 0$ in Theorem 4.

A FORTRAN algorithm for computing $R^+(\alpha;\lambda)$, based on Corollary 4.1, is listed in the appendix.

Notice that the condition $\lambda_j + \cdots + \lambda_k > 0$ is automatically satisfied in the special case $\alpha_2 = \cdots = \alpha_k = 1$, since $\lambda_i = e_i + r_i - r_{i+1}$ and $r_{k+1} = 0$ implies $\lambda_j + \cdots + \lambda_k = e_j + \cdots + e_k + r_j$, which is clearly positive.

5. EXAMPLE

Two examples are given. Both are based on the same set of data kindly supplied by Dr. Robert Reitano of the John Hancock Mutual Life Insurance Company. In the first example, summarized in Table 1, the ages range from 35 to 64. This case was previously analyzed by Broffitt (1984b) using the ordered prior, which provided the values of θ^{IBO} given in Table 1. Table 2 contains the results for the second example which expands the age range to include ages from 30 to 95.

Since for the additive prior we shall assume the special case $a_2 = \cdots = a_k = 1$, it remains to select the prior parameters $a_1, r_1, r_2, \ldots, r_k$. This was done by solving

$$a_j/r_j = \theta_j^0 - \theta_{j-1}^0, \qquad j = 1, \ldots, k \ \ (\theta_0^0 = 0)$$

and

$$a_1/r_1^2 = v_1^0,$$

which results in

$$\begin{aligned} a_1 &= (\theta_1^0)^2/v_1^0 \\ r_1 &= \theta_1^0/v_1^0 \\ r_j &= 1/(\theta_j^0 - \theta_{j-1}^0), \qquad j = 2, \ldots, k. \end{aligned}$$

The prior variance v_1^0 was computed as

$$v_1^0 = \frac{m q_x^0}{e_1(1 - q_x^0)}, \tag{5.1}$$

where $q_x^0 = 1 - \exp(-\theta_1^0)$ and $m = 2$. This may be motivated as follows:

$$\theta_1^M = -\ln(1 - q_x^M),$$

where θ_1^M and q_x^M denote MLE's. Then by the delta method

$$\begin{aligned} \mathrm{Var}(\theta_1^M) &\doteq \left[\left. \frac{d\theta_1^M}{dq_x^M} \right|_{q_x^M = q_x} \right]^2 \mathrm{Var}(q_x^M) \\ &= \frac{\mathrm{Var}(q_x^M)}{(1 - q_x)^2}. \end{aligned}$$

Table 1. *Basic Data and Isotonic Bayes Graduations for Ages 35 to 64.*

Age	d	e	θ^0	θ^M	θ^{IBO}	θ^{IBA}
35	0	1730.0	0.00135	0.00000	0.00105	0.00088
36	4	1902.0	0.00145	0.00210	0.00119	0.00101
37	2	2424.0	0.00157	0.00083	0.00132	0.00112
38	4	2678.0	0.00173	0.00149	0.00147	0.00128
39	3	2617.5	0.00190	0.00115	0.00163	0.00143
40	6	2808.0	0.00211	0.00214	0.00181	0.00163
41	4	3007.0	0.00233	0.00133	0.00200	0.00180
42	12	3371.0	0.00259	0.00356	0.00222	0.00203
43	5	3735.5	0.00288	0.00134	0.00239	0.00218
44	6	3690.0	0.00321	0.00163	0.00261	0.00240
45	7	4145.5	0.00357	0.00169	0.00292	0.00270
46	14	4252.0	0.00397	0.00329	0.00349	0.00332
47	16	4210.0	0.00441	0.00380	0.00416	0.00403
48	31	4611.5	0.00489	0.00672	0.00501	0.00494
49	30	5096.0	0.00541	0.00589	0.00543	0.00535
50	33	6105.5	0.00599	0.00540	0.00581	0.00570
51	36	6483.0	0.00663	0.00555	0.00625	0.00611
52	47	7360.5	0.00736	0.00639	0.00686	0.00668
53	56	8139.0	0.00817	0.00688	0.00764	0.00746
54	87	8789.5	0.00906	0.00990	0.00908	0.00898
55	82	8738.0	0.01003	0.00938	0.00982	0.00969
56	111	9388.5	0.01105	0.01182	0.01076	0.01053
57	91	9275.5	0.01213	0.00981	0.01128	0.01099
58	105	9496.5	0.01329	0.01106	0.01222	0.01182
59	150	9915.0	0.01459	0.01513	0.01404	0.01353
60	124	9541.0	0.01609	0.01300	0.01478	0.01419
61	134	8719.0	0.01790	0.01537	0.01633	0.01560
62	160	9179.0	0.02007	0.01743	0.01848	0.01774
63	216	8777.0	0.02262	0.02461	0.02340	0.02342
64	213	8613.5	0.02549	0.02473	0.02523	0.02504

Table 2. *Basic Data and θ^{IBA} for Ages 30 to 95.*

Age	d	e	θ^0	θ^M	θ^{IBA}
30	1	1559.5	0.00113	0.00064	0.00072
31	2	1547.0	0.00118	0.00129	0.00077
32	0	1641.0	0.00120	0.00000	0.00079
33	1	1800.5	0.00123	0.00056	0.00082
34	3	1680.5	0.00128	0.00179	0.00088
35	0	1730.0	0.00135	0.00000	0.00094
36	4	1902.0	0.00145	0.00210	0.00105
37	2	2424.0	0.00157	0.00083	0.00116
38	4	2678.0	0.00173	0.00149	0.00131
39	3	2617.5	0.00190	0.00115	0.00146
40	6	2808.0	0.00211	0.00214	0.00164
41	4	3007.0	0.00233	0.00133	0.00181
42	12	3371.0	0.00259	0.00356	0.00204
43	5	3735.5	0.00288	0.00134	0.00219
44	6	3690.0	0.00321	0.00163	0.00240
45	7	4145.5	0.00357	0.00169	0.00271
46	14	4252.0	0.00397	0.00329	0.00332
47	16	4210.0	0.00441	0.00380	0.00403
48	31	4611.5	0.00489	0.00672	0.00494
49	30	5096.0	0.00541	0.00589	0.00535
50	33	6105.5	0.00599	0.00540	0.00570
51	36	6483.0	0.00663	0.00555	0.00611
52	47	7360.5	0.00736	0.00639	0.00668
53	56	8139.0	0.00817	0.00688	0.00746
54	87	8789.5	0.00906	0.00990	0.00898
55	82	8738.0	0.01003	0.00938	0.00969
56	111	9388.5	0.01105	0.01182	0.01053
57	91	9275.5	0.01213	0.00981	0.01099
58	105	9496.5	0.01329	0.01106	0.01182
59	150	9915.0	0.01459	0.01513	0.01353
60	124	9541.0	0.01609	0.01300	0.01419
61	134	8719.0	0.01790	0.01537	0.01560
62	160	9179.0	0.02007	0.01743	0.01774

Table 2 (continued)

Age	d	e	θ^0	θ^M	θ^{IBA}
63	216	8777.0	0.02262	0.02461	0.02331
64	213	8613.5	0.02549	0.02473	0.02477
65	153	6110.5	0.02856	0.02504	0.02628
66	177	6032.5	0.03171	0.02934	0.02881
67	175	5710.5	0.03483	0.03065	0.03088
68	173	5443.5	0.03786	0.03178	0.03312
69	193	5136.5	0.04085	0.03757	0.03717
70	208	5054.0	0.04392	0.04116	0.04057
71	203	4558.5	0.04725	0.04453	0.04348
72	193	4415.5	0.05101	0.04371	0.04597
73	223	4254.5	0.05536	0.05242	0.05197
74	211	3631.5	0.06040	0.05810	0.05796
75	226	3479.0	0.06615	0.06496	0.06446
76	213	3035.5	0.07256	0.07017	0.07064
77	202	2603.0	0.07952	0.07760	0.07853
78	204	2216.0	0.08691	0.09206	0.08983
79	196	1953.0	0.09461	0.10036	0.09698
80	184	1722.0	0.10253	0.10685	0.10197
81	135	1389.5	0.11067	0.09716	0.10573
82	128	1224.0	0.11910	0.10458	0.11220
83	139	1076.5	0.12841	0.12912	0.12792
84	145	910.5	0.13999	0.15925	0.14770
85	30	154.0	0.15287	0.19481	0.15764
86	23	163.5	0.16720	0.14067	0.16569
87	21	122.5	0.18303	0.17143	0.17630
88	22	125.0	0.20018	0.17600	0.18800
89	23	93.5	0.21521	0.24599	0.19983
90	14	78.0	0.22685	0.17949	0.20805
91	8	73.0	0.23613	0.10959	0.21563
92	17	57.5	0.24559	0.29565	0.22623
93	6	36.0	0.26146	0.16667	0.24083
94	6	18.0	0.28612	0.33333	0.26910
95	6	17.0	0.31188	0.35294	0.29636

By analogy to the binomial distribution it is reasonable to approximate $\text{Var}(q_x^M)$ by $q_x(1-q_x)/e_1$. Replacing q_x by q_x^0 provides $\text{Var}(\theta_1^M) \doteq q_x^0/[e_1(1-q_x^0)]$. Thus (5.1) is m times the approximate variance of θ_1^M. Choosing $m = 1$ essentially gives equal weight to the observed data and the prior information for estimating θ_1. Larger values of m give less weight to prior information. We chose $m = 2$.

After determining the prior parameters, θ^{IBA} was computed using theorem 3 and corollary 4.1, where $\alpha_j = d_j + a_j$ and $\lambda_j = e_j + r_j - r_{j+1}$, $j = 1, \ldots, k$ with $r_{k+1} = 0$. These values of θ^{IBA} are displayed in Table 1.

Figure 1 shows plots of θ^M, θ^0, $\theta^0 - SD(\Theta)$ and $\theta^0 + SD(\Theta)$, where $SD(\Theta)$ is the standard deviation of the prior distribution of Θ. The one standard deviation band provides an indication of how informative the prior distribution is. Plots of θ^M, θ^0 and θ^{IBA} are displayed in Figure 2. As expected θ^{IBA} provides a compromise between θ^M and θ^0 that is strictly increasing. The estimates θ^{IBO} and θ^{IBA} are compared in Figure 3. They are clearly quite close. The computation of θ^{IBA} was faster since time consuming trial and error calculations were unnecessary in finding the prior parameters. Furthermore, after the prior parameters were determined, it took 173 seconds of CPU time to obtain the values of θ^{IBO}, but only 70 seconds for θ^{IBA}. These calculations were done on a PRIME 750 computer. The smaller CPU time for θ^{IBA} results because the α's are smaller in the additive prior case and consequently $R^+(\cdot;\cdot)$ involves fewer summations than $R(\cdot;\cdot)$.

It is interesting to consider the effect of varying m. Figure 4 contains plots of θ^{IBA} for $m = 1/100$, 1 and 100. In practice it is quite unlikely that values as extreme as 1/100 and 100 would be used, but even in these cases the effects of different m's are negligible after five or ten years.

Finally we consider the second example which considers ages from 30 to 95. This case is included to obtain information on how the Bayesian graduation performs over a very wide range of ages. The prior parameters were determined using the method described above, and the resulting values of θ^{IBA} are listed in Table 2. Figure 5 displays plots of θ^0, θ^M, and θ^{IBA}. Because the values are nearly indistinguishable up to age 80, a separate plot of θ^{IBA} is given in Figure 6.

The CPU time used in calculating these 66 values of θ^{IBA} was 12 minutes and 28 seconds. For a problem of this size, calculation of θ^{IBO} may already be beyond the range of practicality.

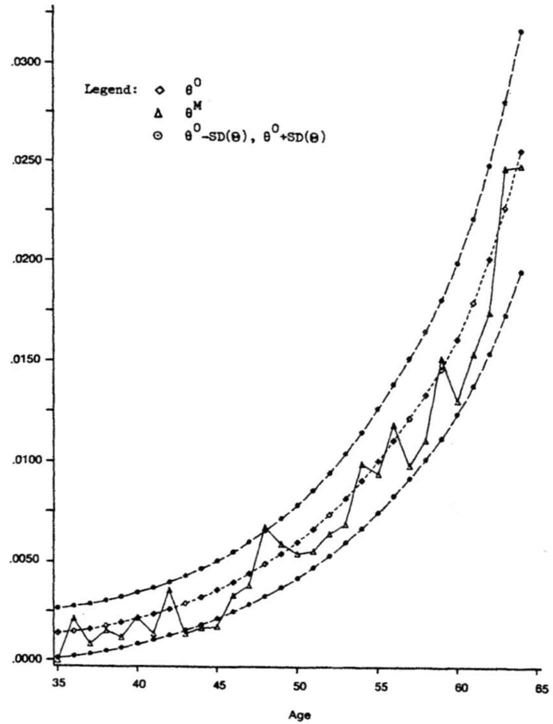

Figure 1. *θ^0, θ^M, $\theta^0 - SD(\Theta)$ and $\theta^0 + SD(\Theta)$ vs. age for ages 35 to 64.*

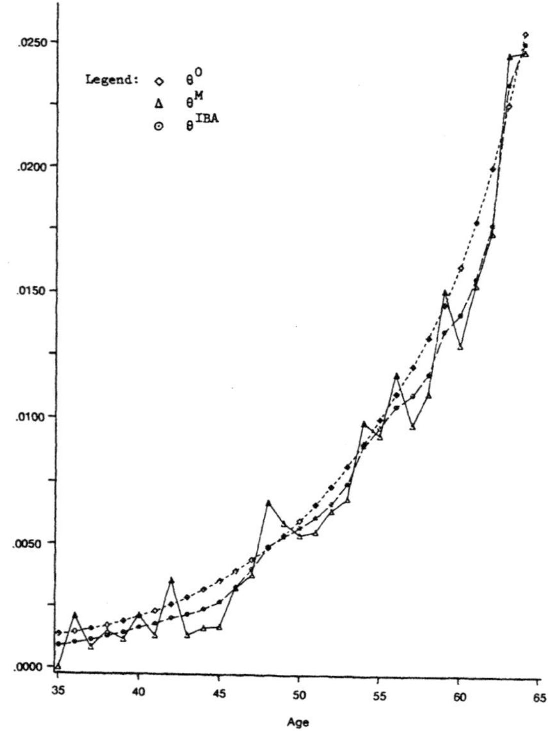

Figure 2. *θ^0, θ^M and θ^{IBA} vs. age for ages 35 to 64.*

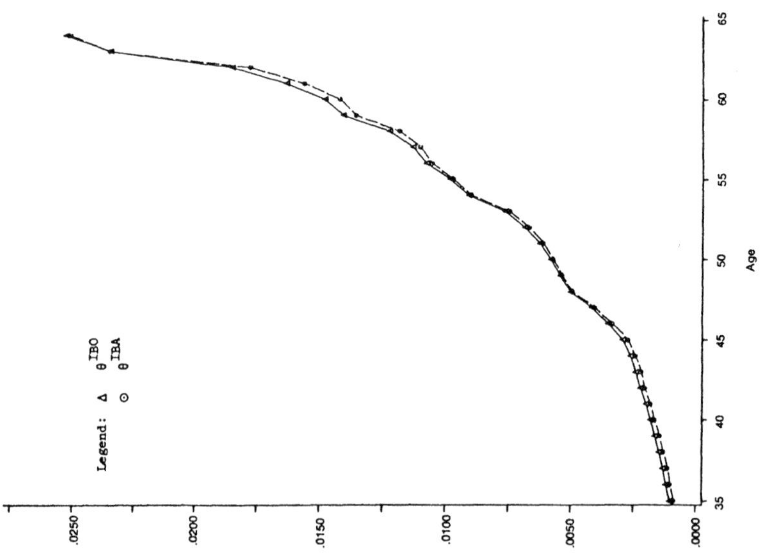

Figure 3. θ^{IBO} *and* θ^{IBA} *vs. age for ages 35 to 65.*

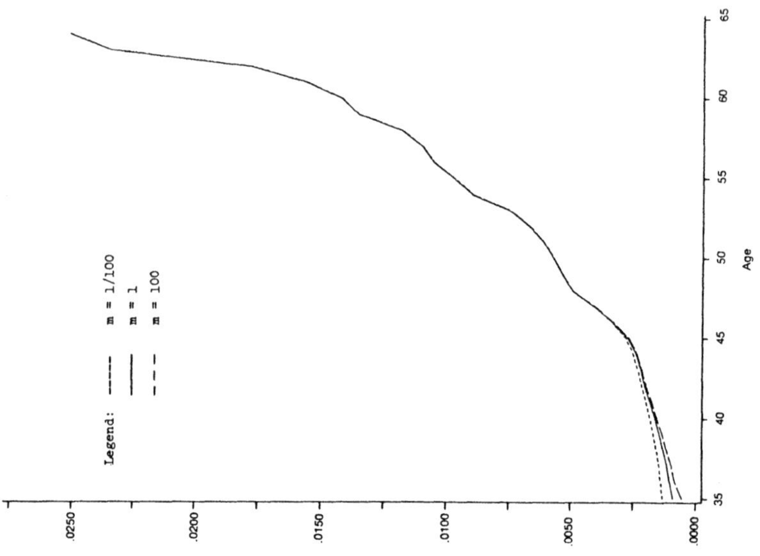

Figure 4. θ^{IBA} *vs. age for* $m = 1/100$, *1 and 100.*

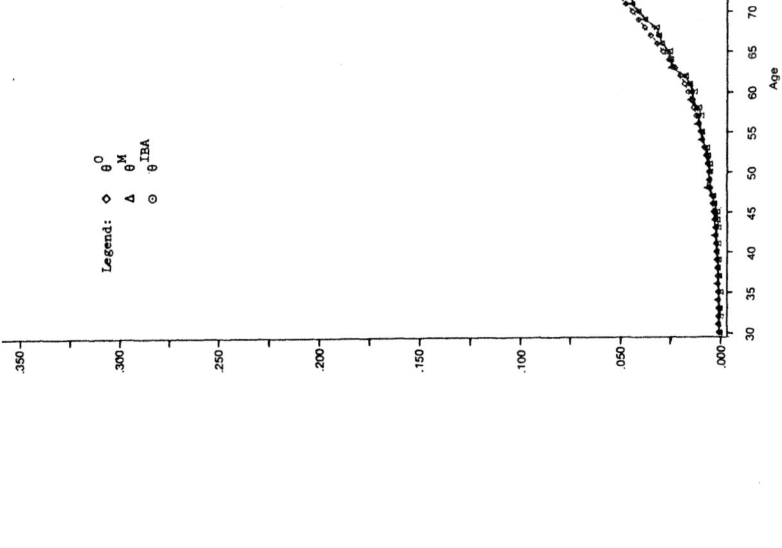

Figure 5. θ^0, θ^M and θ^{IBA} *vs. age for ages 30 to 95.*

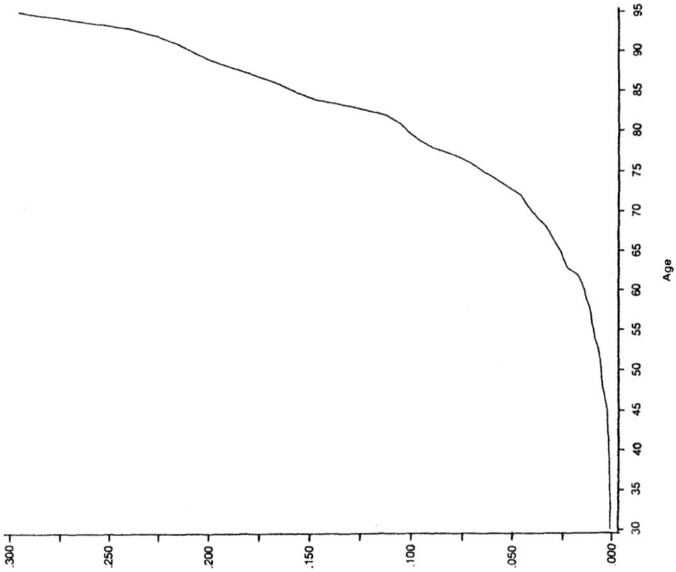

Figure 6. θ^{IBA} *vs. age for ages 30 to 95.*

APPENDIX

```
      SUBROUTINE LGAMMA(K,ALPHA,LAMBDA,RPLUS)
      REAL*8 RPLUS,ALPHA(100),LAMBDA(100),RHO(100),ALFA(100)
      REAL*8 PSL(100),F(100),S(100),ID(100),ARHO(100)
      INTEGER*4 A(100),A2,AN,II
      DO 15 J=1,K
      ALFA(J)=ALPHA(J)
   15 CONTINUE
      DO 5 I=2,K
      A(I)=IDINT(ALFA(I)+.5D0)
    5 ALFA(I)=DBLE(FLOAT(A(I)))
      PSL(K)=LAMBDA(K)
      K1=K-1
      D0 10 I=1,K1
      PSL(K-I)=PSL(K-I+1)+LAMBDA(K-I)
      RHO(K-I)=LAMBDA(K-I)/PSL(K-I)
   10 ARHO(K-I)=DABS(RHO(K-I))
      ARHO(K)=ARHO(1)
      S(1)=1.D0
      DO 20 J=1,K1
      S(J+1)=0.D0
      ID(J)=0.D0
   20 F(J)=1.D0
      A2=A(2)
      DO 30 II=1,A2
      S(2)=ARHO(2)*S(2)
      F(1)=ARHO(2)*F(1)
      S(2)=S(2)+F(1)
      F(1)=F(1)*(1.D0-RHO(1))
      F(1)=F(1)*(ID(1)+ALFA(1))/(ID(1)+1.D0)
   30 ID(1)=ID(1)+1.D0
      IF(K .EQ. 2) GO TO 70
      DO 40 N=3,K
      AN=A(N)
      DO 50 II=1,AN
      S(N)=ARHO(N)*S(N)
      F(N-1)=ARHO(N)*F(N-1)
      S(N)=S(N)+F(N-1)*S(N-1)
      F(N-1)=F(N-1)*(1-RHO(N-1))
      F(N-1)=F(N-1)*(ID(N-1)+ALFA(N-1))/(ID(N-1)+1.D0)
      ID(N-1)=ID(N-1)+1.D0
```

```
   IF(II .EQ. AN) GO TO 40
   N1=N-1
   D0 60 J=2,N1
   S(J)=S(J)+F(J-1)*S(J-1)
   F(J-1)=F(J-1)*(1-RHO(J-1))
   F(J-1)=F(J-1)*(ID(J-1)+ALFA(J-1))/(ID(J-1)+1.D0)
   ID(J-1)=ID(J-1)+1.D0
60 CONTINUE
50 CONTINUE
40 CONTINUE
70 RPLUS=S(K)*(ARHO(K)**(ALFA(1)-ALFA(K)))
   RETURN
   END
```

REFERENCES

Broffitt, J. D. (1984a), "Maximum likelihood alternatives to actuarial estimators of mortality rates". *Transactions of the Society of Actuaries* **36**, 77–122.

Broffitt, J. D. (1984b), "A Bayes estimator for ordered parameters and isotonic Bayesian graduation". *Scandinavian Actuarial Journal*, 231–247.

Chan, L. K., and H. H. Panjer (1983), "A statistical approach to graduation by mathematical formula". *Insurance: Mathematics and Economics* **2**, 33–47.

Steelman, J. M. (1968), "Statistical approaches to mortality estimation". M.S. Thesis, University of Iowa.

Esther Portnoy [1]

BOOTSTRAPPING A GRADUATION

1. INTRODUCTION

The bootstrap is one of several *resampling* methods that have been developed recently by statisticians. Bradley Efron (1982, pp. 2–3), who developed the bootstrap, notes that these methods "are prodigious computational spendthrifts.... An important theme...is the substitution of computational power for theoretical analysis. The payoff...is freedom from the constraints of traditional parametric theory, with its over-reliance on a small set of standard models for which theoretical solutions are available."

The basic idea of the bootstrap is very simple. It begins with a random sample $X_1, \ldots, X_n$ from a unknown distribution F. The sample defines an empirical distribution function $\hat{F}$ which (roughly) approximates F. Next we have some functional $R(X_1, \ldots, X_n, F)$—the sample mean, perhaps, or a set of graduated mortality rates. R may depend explicitly on F, for example via parameters such as its mean or variance. R is itself a random variable, and along with the value it takes on for a particular sample we would generally like to have some idea of its distribution.

The bootstrap idea consists of drawing a pseudo-sample $X_1^*, \ldots, X_n^*$ from the empirical distribution $\hat{F}$ and then estimating the distribution of $R(X_1, \ldots, X_n, F)$ by the distribution of $R^* = R(X_1^*, \ldots, X_n^*, \hat{F})$. Since we know the empirical distribution, we should be able (in theory) to find the distribution of R^*, either by direct calculation or by Monte Carlo approximation.

To illustrate, suppose that R is the sample mean, $\frac{1}{n}\sum X_i$. Then $\mathrm{Var}(R) = \frac{1}{n}\,\mathrm{Var}(X)$; if we do not know $\mathrm{Var}(X)$ we may approximate it by the sample variance, thus obtaining $\mathrm{Var}(R) \doteq \frac{1}{n(n-1)}\sum(x_i - \bar{x})^2$. The bootstrap estimate, on the other hand, is $\mathrm{Var}(R) \doteq \mathrm{Var}(R^*) = (\frac{1}{n})^2\sum(x_i - \bar{x})^2$. This is close, but not exactly equal, to the earlier estimate. The great ad-

[1] Department of Mathematics, University of Illinois at Urbana-Champaign, Urbana, Illinois 61801

I. B. MacNeill and G. J. Umphrey (eds.), Actuarial Science, 41–51.

vantage of the bootstrap is that it generalizes readily to more complicated situations.

If R is quite complicated, then even if we know the distribution F, we might not be able to calculate $\mathrm{Var}(R)$ directly, but we might instead be able to approximate it by a Monte Carlo method. The bootstrap proceeds in much the same way, substituting "pseudo-samples" $X_1^*, \ldots, X_n^*$ for the samples of the usual Monte Carlo. The difficulty comes in deciding whether the results of such resampling are valid and useful.

Efron (1982) has proved a fairly general convergence theorem for the multinomial distribution, though it is not clear whether this is directly applicable to the multidimensional function treated below. Bickel and Freedman (1981) have shown convergence in a number of other situations, and Beran (1982) has proven optimality of the bootstrap in the sense of the rate of convergence of the bootstrap distribution to the true distribution. The purpose of this paper is not to prove any theorem but to demonstrate the method in a particular actuarial application, and to present an example that suggests the shape a theorem might take.

2. AN ACTUARIAL APPLICATION

The particular question to which I want to apply the bootstrap method concerns the significance of observed crossovers in sex-distinct mortality rates. Figure 1 and Table 1 show some mortality rates among insured Medicare enrollees born between 1872 and 1875; the data is from Bayo and Faber (1983). After smoothing by a Whittaker method, the two curves cross at about age 101.5.

One way to estimate the significance of the crossover is to construct a band about each of the smoothed curves, such that we have some reasonable confidence that the "true" mortality rates lie within the bands. If the bands are quite narrow, they will overlap in a small region, roughly a parallelogram, and then we can say with some confidence that the mortality rates do cross, within the range of ages covered by the overlap. But as the bands widen, the overlap increases and so too does the interval for the crossover age. For bands of moderate width the interval for the crossover age covers the whole range of ages in the data, and the significance of the crossover dissipates. Thus, it is very important to know how wide the bands should be. This is the sort of problem for which the bootstrap was designed: estimating some measure of spread for a complicated statistic, when the underlying distribution is unknown.

In order to develop and test a method for this problem, we formulate a similar situation, a multinomial distribution with parameters (hypothet-

Table 1. *Raw and Graduated Mortality Rates Among Insured Medicare Enrollees* (from Bayo and Faber, 1983)

Exact	Number of Deaths		Exposure		Ungraduated Death Rate		Graduated Death Rate	
Age	Male	Female	Male	Female	Male	Female	Male	Female
95.5	4,346	4,500	14,494	17,014	.29985	.26449	.29837	.26635
96.5	3,240	3,545	10,148	12,514	.31927	.28328	.31071	.28134
97.5	2,141	2,660	6,908	8,969	.30993	.29658	.32259	.29682
98.5	1,621	2,020	4,767	6,309	.34005	.32018	.33373	.31272
99.5	1,097	1,399	3,146	4,289	.34870	.32618	.34381	.32900
100.5	730	1,000	2,049	2,890	.35627	.34602	.35253	.34568
101.5	490	640	1,319	1,890	.37149	.33862	.35969	.36283
102.5	275	478	829	1,250	.33172	.38240	.36513	.38045
103.5	192	319	554	772	.34657	.41321	.36875	.39849
104.5	154	219	362	453	.42541	.48344	.37045	.41688
105.5	92	86	208	234	.44231	.36752	.37016	.43558
106.5	24	64	116	148	.20690	.43243	.36784	.45456

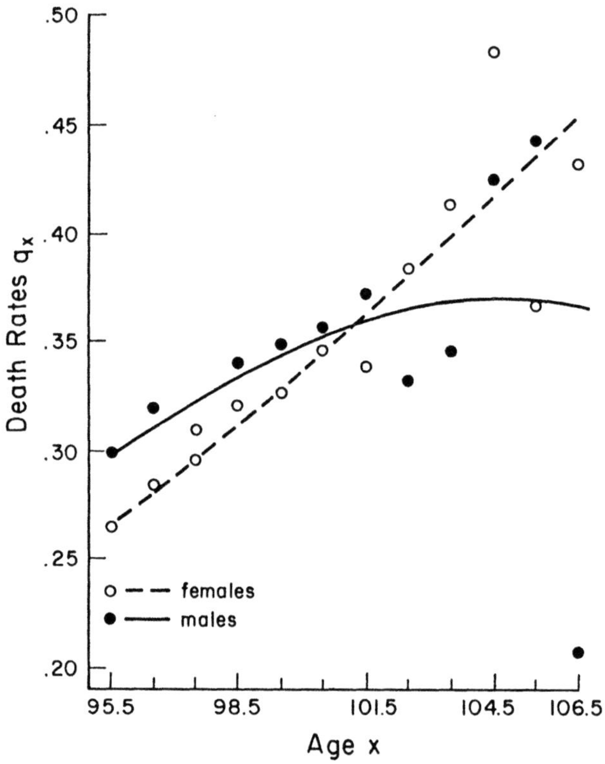

Figure 1. *Mortality rates among insured Medicare Enrollees (Data from Bayo and Faber, 1983).*

ical mortality rates) given in Table 2. Realism is not a consideration in choosing these rates; rather, we want a fairly substantial change over a short range, and a smooth progression. These rates actually lie on a cubic, and no graduation method (even linear regression) will change them very much.

Now we carry out 200 Monte Carlo simulations of the following experiment: assume 1000 lives at age 1. Generate Poisson random variables (approximating the binomial) for the number of deaths in each succeeding year, based on the rates given by the cubic law, and the number of survivors $\hat{n}_x$ from the preceding year. This generally leaves about 900 survivors at the end of the 10-year period. Now do a Whittaker graduation of the observed "mortality" rates ($z = 3, h = 100,000, w_x = \hat{n}_x/q_x(1 - q_x)$ using q_x from

Table 2. *Hypothetical Mortality Rates for Illustrative Example*

x	$1000q_x$	95% "confidence" band
1	3.850	1.077– 7.360
2	5.050	1.875– 9.071
3	6.330	2.782–10.844
4	7.770	3.831–12.757
5	9.331	5.015–14.796
6	11.040	6.345–16.984
7	12.910	7.833–19.338
8	14.960	9.494–21.879
9	17.211	11.349–24.632
10	19.670	13.403–27.604

Table 2), and note how the 200 smoothed sequences thus obtained group themselves around the original cubic.

Several things appear on examination of the results. As has been noted before (e.g., Kimeldorf and Jones, 1967), the Whittaker method does have some end-value problems. The spread of values was considerably larger at $x = 1$ and $x = 10$ than at the intermediate points; in fact, occasionally the graduated value of q_1 was negative. For $x = 2, \ldots, 9$ the spread was roughly proportional to $\sqrt{q_x}$, and slightly greater above q_x than below. This is as expected, because the distribution of the binomial proportion with small q_x is skewed and has variance $q_x(1 - q_x)/\hat{n}_x$, here approximately $q_x/1000$. On the basis of these observations, a band was chosen in the form $(q_x - R_1\sqrt{q_x}$, $q_x + R_2\sqrt{q_x})$, so that 2.5% of the runs (5 in 200) produced graduated values greater than $q_x + R_2\sqrt{q_x}$ for some x between 2 and 9, and another 2.5% similarly dipped below the band. (None of the 200 runs gave values both above and below the band.) For this 200-run sample, we find $R_1 = 1.413/\sqrt{1000}$ and $R_2 = 1.789/\sqrt{1000}$, compared to an expected spread of about $2/\sqrt{1000}$ for simple binomial sampling. Thus, we see that the graduation has (as one would hope) decreased the spread.

What has been done so far is not bootstrapping but an ordinary Monte Carlo method. The bootstrap enters when we consider one of the Monte

Carlo runs, set aside the hypothetical cubic rates that underlie it, and try to find a confidence band about the smoothed Monte Carlo rates.

The program is the same as before, with two important exceptions. First, the sample (or "pseudo-sample") is drawn from a multinomial distribution whose parameters are determined not by the hypothetical cubic rates q_x but by the "observed" (here, Monte Carlo) rates $\hat{q}_x$. This is the essence of the bootstrap. Second, in determining the weights $\hat{n}_x/q'_x(1-q'_x)$ to be used in the Whittaker graduation, we cannot take q'_x equal to the (supposedly unknown) q_x. Nor are the $\hat{q}_x$ an acceptable substitute, since some of them may vanish. (With $n_1 = 1000$ and $q_1 = .00385$, the probability that $\hat{q}_1 = 0$ is about 2%.) Besides, if some $\hat{q}_x$ is an obvious outlier lower than adjacent values, then taking $q'_x = \hat{q}_x$ would assign more weight to precisely that point that we suspect is not very close to the true value.

We need a smoother sequence of rates to be used only for determining the weights. Almost any *ad hoc* method, such as a graphic graduation, should be acceptable. In this testing situation, since the process is to be repeated with several hundred different sets of "observed" rates, it is essential that the initial smoothing be automatic. Noting the near-linearity of the hypothetical rates, I used simple linear regression on the $\hat{q}_x$ to produce the q'_x. One drawback was that about 2% of the time this produced $q'_1 < 0$; pending further study, these runs were set aside.

For each of 200 sets of simulated $\hat{q}_x$ (not the same ones used in the initial determination of the band around the hypothetical rates), a confidence band was constructed by the following procedure:

(1) Determine q'_x by linear regression on $\hat{q}_x$.

(2) Smooth $\hat{q}_x$ by a Whittaker graduation with $z = 3, h = 100{,}000$, $w_x = \hat{n}_x/q'_x(1-q'_x)$, to obtain rates $\hat{q}'_x$.

(3) Resample from a multinomial distribution with parameters determined by $\hat{q}_x$, obtaining rates q^*_x.

(4) Smooth q^*_x as in step 2.

(5) Repeat steps 3 and 4, 100 times.

(6) Determine a band of the form $(\hat{q}'_x - R_1\sqrt{\hat{q}'_x}, \hat{q}'_x + R_2\sqrt{\hat{q}'_x})$ that covers 95% of the smoothed, resampled rates.

One example of a Monte Carlo run and the resulting $\hat{q}'_x$ and band are illustrated in Figure 2.

The intention of this procedure was to produce a band that, in about 95% of cases, will include the true rates underlying the $\hat{q}_x$. Of the 200 repetitions above, six were discarded because some q'_x or $\hat{q}'_x$ was negative. Of the remaining 194, 187 or 96.4% covered the hypothetical cubic rates. Thus, the method proposed seems to do about what we hoped: the band

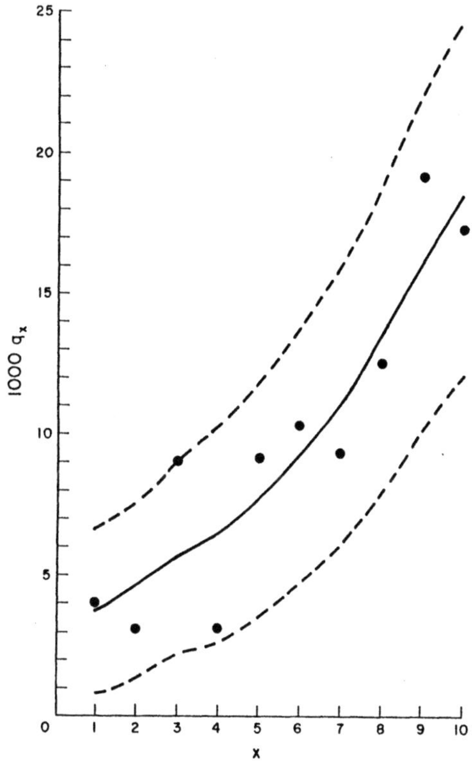

Figure 2. *"Confidence" band derived from a Monte Carlo simulation of hypothetical mortality rates.*

will usually, but not always, include the true values. (Certainly the 100 repetitions in step 5 are too few for real accuracy, and in the example below it will be necessary to repeat many more times; but this should suffice for purposes of illustration.)

Finally, let us return to the actual problem that led to this study, the crossover question. Now we really don't know the true mortality rates. To develop a band, I resampled 4000 times from each of the empirical distributions (multinomial with parameters determined by the male resp. female mortality rates), and smoothed using $z = 3$, $h = 50,000$, $w_x = \hat{n}_x$. (Bayo and Faber used exposures as weights. Some preliminary testing with weights $\hat{n}_x/q'_x(1-q'_x)$ indicated the results would be changed very little.)

In this example the exposures drop off quite dramatically, from about

15,000 at age 95.5 to fewer than 150 at age 106.5. Also, the approximation $1 - \hat{q}'_x \doteq 1$ is not adequate. Consequently the band width was taken proportional to $\sqrt{q'_x(1-q'_x)/n_x}$, where the q'_x are the graduated rates and n_x the exposures from Table 1.

As in the earlier example, the band width is determined by two parameters,

$$r_1 = \max_{96.5 \leq x \leq 105.5} \frac{\hat{q}'_x - q'^*_x}{\sqrt{q'_x(1-q'_x)/n_x}}$$

and

$$r_2 = \max_{96.5 \leq x \leq 105.5} \frac{q'^*_x - \hat{q}'_x}{\sqrt{q'_x(1-q'_x)/n_x}}.$$

Some percentiles for these two parameters are given in Table 3. The rates $\hat{q}'_x$ differ from Bayo and Faber's q'_x because here the Whittaker smoothing was applied only over the interval from 95.5 to 106.5. The $\hat{q}'_x$, and bands covering 90% and 95% of the Monte Carlo runs, are given in Table 4, and the 95% bands illustrated in Figure 3.

Table 3. *Percentiles for Band-Width Parameters r_1 and r_2*

	Males		Females	
	r_1	r_2	r_1	r_2
95%	1.11176	1.18064	1.13011	1.16657
95.75%	1.14419	1.21449	1.15727	1.21628
97.5%	1.25662	1.32889	1.25160	1.34236

It is clear that mortality rates for females are significantly below those for males at ages up to 98.5, after which the bands overlap. The band about the female rates almost emerges above the male band at age 106.5, but not quite, when we use 97.5% critical values of r_1 and r_2. The bands derived from the 95% values (thus, the bands that cover about 90% of the Monte Carlo runs) do just barely cross. The conclusion that the crossover is just barely significant agrees with a different test applied to this data in Portnoy (1985).

Table 4. *Graduated Mortality Rates and "Confidence" Bands for Insured Medicare Enrollees*

Males: Age	$\hat{q}'_x$	90% band	95% band
95.5	.30078	.29654–.30528	.29599–.30584
96.5	.31239	.30728–.31782	.30661–.31851
97.5	.32359	.31734–.33024	.31652–.33107
98.5	.33465	.32706–.34272	.32607–.34374
99.5	.34499	.33556–.35499	.33433–.35625
100.5	.35390	.34216–.36637	.34063–.36794
101.5	.36087	.34617–.37648	.34425–.37844
102.5	.36551	.34692–.38526	.34449–.38774
103.5	.36753	.34476–.39172	.34179–.39475
104.5	.36636	.33820–.39626	.33454–.40001
105.5	.36131	.32427–.40064	.31945–.40557
106.5	.35205	.30228–.40490	.29580–.41153
Females:			
95.5	.26510	.26128–.26905	.26087–.26964
96.5	.28219	.27764–.28688	.27715–.28759
97.5	.29839	.29293–.30403	.29235–.30488
98.5	.31393	.30732–.32074	.30662–.32177
99.5	.32914	.32103–.33751	.32016–.33877
100.5	.34479	.33480–.35511	.33373–.35666
101.5	.36151	.34916–.37426	.34768–.37634
102.5	.37953	.36402–.39554	.36235–.39795
103.5	.39834	.37843–.41890	.37629–.42199
104.5	.41722	.39103–.44424	.38822–.44832
105.5	.43564	.39901–.47346	.39507–.47916
106.5	.45357	.40731–.50132	.40233–.50851

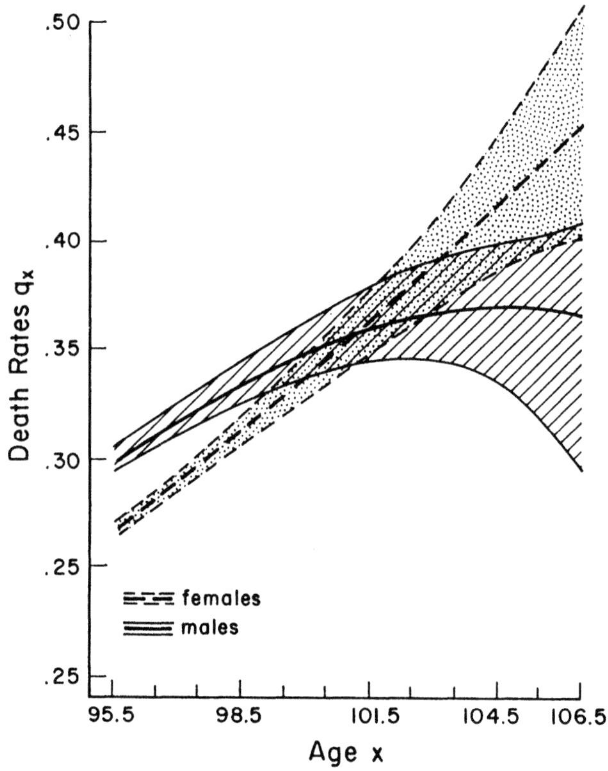

Figure 3. *95% "confidence" bands about graduated mortality rates among insured Medicare enrollees.*

The percentiles given in Table 3 are, of course, only estimates of parameters of the true (and unknown) distributions of r_1 and r_2. An indication of their accuracy can be obtained by noting that, if ν were the true 95th percentile of r_1, then the probability that, in 4000 trials, fewer than 170 exceed ν is approximately .02 (the expected number exceeding ν is 200, and its variance is 190). A conservative estimate for the 90% band could be based on the 95.75% points for r_1 and r_2; these two bands just barely overlap at 106.5.

3. QUESTIONS FOR FURTHER STUDY

At several points, the analysis in this paper has been *ad hoc*, adapted to the particular nature of the data set being treated. The criterion used for determining the "confidence" band constitutes one example; the difficulty with end values is particularly unfortunate, and some method should be sought to handle them more satisfactorily. The effect of the smoothing constant h is also of interest. Intuitively, one expects that a larger value of h should result in narrower bands, since the smoothed values are less constrained to lie near the resampled values.

A study like the one presented here for the hypothetical cubic law should be done for a hypothetical law that is much less smooth, where the Whittaker method may in fact oversmooth observed values. Other graduation methods may also be analyzed via the bootstrap, which seems to offer some advantages over other methods of obtaining confidence intervals.

Another question that should be addressed in more generality is how many bootstrap iterations are necessary to obtain reliable results. The search for confidence intervals involves estimating the tails of distributions (here, of r_1 and r_2) and, therefore, requires more iterations than estimates of more centrally located parameters.

REFERENCES

Bayo, F. R., and J. R. Faber (1983), "Mortality experience around age 100." *Transactions of the Society of Actuaries* **35**, 37–59.

Beran, R. (1982), "Estimated sampling distributions: the bootstrap and competitors." *Annals of Statistics* **10**, 212–225.

Bickel, P. J., and D. A. Freedman (1981), "Some asymptotic theory for the bootstrap." *Annals of Statistics* **9**, 1196–1217.

Efron, B. (1982), *The Jackknife, the Bootstrap, and Other Resampling Plans.* Philadelphia: SIAM.

Kimeldorf, G. S., and D. A. Jones (1967), "Bayesian graduation." *Transactions of the Society of Actuaries* **19**, 66–112.

Portnoy, Esther (1985), "Crossover in mortality rates by sex." *Transactions of the Society of Actuaries* **38**, to appear.

Hans U. Gerber [1]

ACTUARIAL APPLICATIONS OF UTILITY FUNCTIONS

1. WHY UTILITY FUNCTIONS?

Certain phenomena cannot be explained by looking at expected values only. Then often an answer can be found, or a given behaviour can be explained, by assuming utility functions.

We all recall the Petersburg paradox of Bernoulli. An actuarial version is the following problem: Let S be the claims to be paid by an insurer; S is a random variable with a supposedly known distribution. What is an equitable premium π for S? We are caught in a dilemma: on the one hand "equitable" means equality of expectations, on the other hand the insurer would certainly want π to be greater than $E(S)$.

The philosophy is not to claim that economical agents have a utility function. Instead, we deliberately assume that they have a utility function, and discuss the resulting conclusions, which often could not be obtained otherwise. Thus a utility function plays the role of a deus from my china (ex machina)!

2. WHAT IS A UTILITY FUNCTION?

For our purposes a utility function is simply a function $u(x)$ with $u'(x) > 0$ and $u''(x) < 0$. The interpretation is that x is the fortune of an economic agent (measured in dollars), and u its value in some sense. Thus u is an increasing function (obviously: the more the better) that is concave; the second property means that a dollar handed out to a millionaire means less to him than a dollar given to a poor man.

If we replace a utility function by one of its linear transformations, the conclusions stay the same. In fact, the function of importance is the associ-

[1] Ecole des H.E.C., Université de Lausanne, 1015 Lausanne, Switzerland

I. B. MacNeill and G. J. Umphrey (eds.), Actuarial Science, 53–61.

ated risk aversion function

$$r(x) = \frac{-u''(x)}{u'(x)} = -\frac{d}{dx}\ln u'(x), \tag{1}$$

which is invariant under linear transformations of the utility function. Note that the risk aversion function is positive according to our assumptions.

Let us return to the problem of defining an equitable premium. If we assume that the insurer has a utility function u, and stipulate that the premium is to be fair in terms of utility, π is obtained as the solution of the equation

$$E\left[u(x+\pi-S)\right] = u(x). \tag{2}$$

Thus the expected utility with the new contract should be equal to the utility without the new contract. From Jensen's inequality we see that the resulting π is indeed greater than $E[S]$.

In general equation (2) cannot be solved explicitly for π. If, however, the distribution of S is closely concentrated around its mean, π can be approximated as follows:

$$\pi \approx E(S) + \frac{1}{2}r(x)\mathrm{Var}(S). \tag{3}$$

For a precise meaning and a generalization of this statement see Gerber (1985).

The exponential utility function,

$$u(x) = \frac{1}{a}(1-e^{-ax}), \quad a > 0, \tag{4}$$

plays a central role. On the one hand it is simple, since $r(x) = a$ is constant. On the other hand many calculations can be done explicitly, if exponential utility functions are assumed. For example, equation (2) can be solved for π; one finds that

$$\pi = \frac{1}{a}\ln E\left[e^{aS}\right]. \tag{5}$$

3. RISK EXCHANGE BETWEEN INSURANCE COMPANIES

In this model we consider n companies, labelled from 1 to n. Let X_i be the surplus of company i at the end of the year. Typically, X_i is the surplus at the beginning of the year plus the premiums received minus the claims to be paid. Mostly because of the latter, X_i is a random variable. We assume that the joint distribution of $(X_1, \cdots, X_n)$ is known.

By trading, or exchanging risks, the companies can try to improve their situation in some sense. The result of such an exchange is a random vector $(Y_1, \cdots, Y_n)$, where Y_i is the modified surplus of company i. The only restriction is that the combined surplus after such an exchange is equal to the combined surplus before:

$$Y_1 + Y_2 + \cdots + Y_n = X_1 + X_2 + \cdots + X_n. \tag{6}$$

In the following we shall denote $X_1 + \cdots + X_n$ by X.

A particularly simple type of exchange is given by the formula

$$Y_i = q_i X + c_i; \tag{7}$$

here q_i is the quota of company i $(q_1 + \cdots + q_n = 1)$, and c_i is the deterministic side payment received by company i $(c_1 + \cdots + c_n = 0)$.

4. PARETO OPTIMAL EXCHANGES

We shall now assume that company i uses a utility function u_i to evaluate an exchange. Thus the value of $(Y_1, \cdots, Y_n)$ to company i is $E[u_i(Y_i)]$, $i = 1, \ldots, n$.

An exchange is said to be Pareto optimal, if there is no other exchange that is better for all companies. For geometric reasons the family of Pareto optimal exchanges is identical with the solutions of the following problems: Choose positive constants $k_1, \cdots, k_n$ and maximize

$$k_1 E[u_1(Y_1)] + \cdots + k_n E[u_n(Y_n)]. \tag{8}$$

Using variational calculus and the concavity of the utility functions we arrive at the following equivalent condition:

$$k_i u_i'(Y_i) \text{ is independent of } i, \ i = 1, \cdots, n. \tag{9}$$

In the literature this fact is known as the theorem of Borch; see Bühlmann (1970).

Note that (9) amounts to $n - 1$ independent conditions. Together with (6) we can determine the Pareto optimal exchange $(Y_1, \cdots, Y_n)$ at least in principle. If we assume exponential utilities,

$$u_i(x) = \frac{1}{a_i}(1 - e^{-a_i x}), \ a_i > 0, \tag{10}$$

we find an exchange of the form (7), with

$$q_i = \frac{a}{a_i}, \text{ where } \frac{1}{a} = \frac{1}{a_1} + \cdots + \frac{1}{a_n}, \tag{11}$$

$$c_i = \frac{1}{a_i}(\ln k_i - \sum_{j=1}^{n} \frac{a}{a_j} \ln k_j). \tag{12}$$

Thus the quotas are the same for all Pareto optimal risk exchanges, which differ only by the side payments.

In general there is an $(n-1)$-parametric family of Pareto optimal risk exchanges (a natural parametrization is given by the vector of the k_i's, if we impose the condition that their sum is 1, for example). Some of these cannot be realized, because one or several companies would be worse off with the exchange than without. But even the set of Pareto optimal risk exchanges that improve the expected utilities of all companies is still $(n-1)$-dimensional, and the actual exchange has to be found by some kind of a bargaining process. In the next section we shall refine the model to get more definite answers.

5. EQUILIBRIUM (BÜHLMANN, 1980)

We assume that the n companies of Section 4 form a pool. Company i can buy a payment or reimbursement R_i (a random variable) from the pool for a premium of π_i, $i = 1, \cdots, n$. Note that we can interpret such an arrangement as a risk exchange with

$$Y_i = X_i - \pi_i + R_i \tag{13}$$

as long as

$$R_1 + R_2 + \cdots + R_n = \pi_1 + \cdots + \pi_n. \tag{14}$$

The premiums are determined according to a premium intensity P; this is a random variable with $P > 0$ and $E(P) = 1$. Then

$$\pi_i = E[PR_i] = E[R_i] + \text{Cov}(P, R_i). \tag{15}$$

Admittedly this is a very simple way to calculate the premiums, yet it allows the premium to depend not only on the distribution of the payment, but also on its relation to other random variables of interest, such as $X_1, \ldots, X_n$. Typically P is a function of $X_1, \ldots, X_n$.

Company i wants to choose R_i in order to

$$\text{maximize} \quad E\left[u_i(X_i - \pi_i + R_i)\right], \tag{16}$$

where $\pi_i = E(PR_i)$. Of course the optimal R_i is determined only up to an additive constant (if a constant is added to a reimbursement R_i, the same constant is added to the corresponding premium π_i, and their difference remains unchanged). The optimal R_i is obtained from the condition that

$$\frac{u_i'(X_i - \pi_i + R_i)}{E\left[u_i'(X_i - \pi_i + R_i)\right]} = P. \tag{17}$$

We call $(P, R_1, R_2, \ldots, R_n)$ an equilibrium if (14) holds, and (17) is satisfied for $i = 1, 2, \ldots, n$.

The risk exchange that results from an equilibrium is Pareto optimal. This follows from the fact that (17) implies (9), if we set

$$k_i = 1/E[u_i'(X_i - \pi_i + R_i)].$$

In the case of exponential utility functions, see (10), condition (17) gives

$$R_i = -X_i - \frac{1}{a_i} \ln P + d_i, \tag{18}$$

where d_i is an arbitrary constant. From this, (14) and the condition that $E(P) = 1$ we obtain

$$P = \frac{e^{-aX}}{E\left[e^{-aX}\right]}. \tag{19}$$

If we choose d_i so that $E(R_i) = 0$, using (19) we get

$$R_i = -(X_i - \mu_i) + q_i(X - \mu), \tag{20}$$

where $\mu_i = E(X_i)$, $\mu = E(X)$, and $q_i = a/a_i$ as in (11). Then

$$\pi_i = -\text{Cov}(P, X_i) + q_i \text{Cov}(P, X) \tag{21}$$

and

$$Y_i = q_i X - \pi_i + \mu_i - q_i \mu. \tag{22}$$

This last expression shows that the resulting risk exchange is indeed Pareto optimal.

If we assume that $a_1, \ldots, a_n$ are small, the following first order approximations are obtained:

$$P \approx 1 - a(X - \mu), \tag{19'}$$

$$\pi_i \approx a\mathrm{Cov}(X, X_i) - aq_i \mathrm{Var}(X). \tag{21'}$$

6. THE BOWLEY SOLUTION (CHAN AND GERBER, 1985)

For the case of two companies we shall now discuss an alternative solution. Since the two companies will play asymmetric roles, we shall call company 1 the "first insurer" and company 2 the "reinsurer"; these designations should be interpreted with a grain of salt.

The first insurer can buy a payment of $R = R_1$ for a premium of $\pi = \pi_1$, which is calculated according to the rule $\pi = E[PR]$. For given premium intensity P, the first insurer chooses R to maximize his expected utility; thus, as in (17), R is obtained from the condition that

$$\frac{u_1'(X_1 - \pi + R)}{E\left[u_1'(X_1 - \pi + R)\right]} = P. \tag{23}$$

This defines the demand for reinsurance, i.e., the payment R that the first insurer wants to buy as a function of the price intensity P. We assume that the reinsurer knows this function and has the monopoly. Thus he will choose P to maximize his expected utility:

$$E\left[u_2(X_2 + \pi - R)\right]. \tag{24}$$

In this expression R depends on P as indicated in (23). The resulting pair (P, R) is called the Bowley solution.

Here we shall discuss the most simple case, where

$$u_1(x) = \frac{1}{\alpha}(1 - e^{-\alpha x}),\ u_2(x) = x, \tag{25}$$

i.e., where the first insurer has a constant risk aversion $\alpha > 0$ and the reinsurer is risk neutral. Thus the first insurer's demand function is

$$R = -X_1 - \frac{1}{\alpha}\ln P + d, \tag{26}$$

where d is an arbitrary constant, see (18), and the reinsurer chooses P in order to maximize his expected gain:

$$\pi - E(R) = -E(PX_1) - \frac{1}{\alpha}E(P\ln P) + E(X_1) + \frac{1}{\alpha}E(\ln P). \tag{27}$$

Using variational calculus one finds that P is obtained from the equation

$$-X_1 - \frac{1}{\alpha}\ln P + \frac{1}{\alpha P} = c, \tag{28}$$

where the constant c must be chosen such that $E(P) = 1$. Thus, the Bowley solution under assumptions (25) is given by formulas (28) and (26). Furthermore, from (27) and (28) one can show that the reinsurer's expected gain is

$$\pi - E(R) = \frac{1}{\alpha}\left(E\left(\frac{1}{P}\right) - 1\right), \tag{29}$$

which is positive because of Jensen's inequality.

It is interesting to note that for the utility functions (25) the equilibrium of Section 5 (with $n = 2$) does not provide a satisfactory solution. To see this, notice that (17) with $i = 2$ implies that $P \equiv 1$, so that $\pi_i = E(R_i)$, which, together with (17) for $i = 1$ implies $Y_1 = \mu_1$. Then from (6) we obtain

$$Y_2 = X_1 + X_2 - \mu_1. \tag{30}$$

Thus $E[u_2(Y_2)] = E[u_2(X_2)]$, which means that the reinsurer would not gain from the arrangement.

7. OPTIMAL PURCHASE OF REINSURANCE (DEPREZ AND GERBER, 1985)

We shall assume that reinsurance premiums are calculated according to a more general principle than $\pi = E(PR)$. Thus let $\pi = H(R)$ denote the premium that the first insurer has to pay for a reimbursement of R; mathematically, H is a functional over a set of random variables.

The problem is to choose R to maximize

$$E[u_1(X_1 - H(R) + R)]. \tag{31}$$

If H is convex and sufficiently regular, the optimal R is obtained as the solution of the equation

$$\frac{u_1'(X_1 - H(R) + R)}{E[u_1'(X_1 - H(R) + R)]} = H'(R). \tag{32}$$

Here $H'(R)$ denotes the gradient of H at R; this is the random variable for which

$$\frac{d}{dt}H(R + tQ)\,|_{t=0} = E[QH'(R)] \quad \text{for all } Q. \tag{33}$$

If $H(R) = E(PR)$, $H'(R) = P$. Thus (32) is a generalization of (17) and (23).

Suppose now that

$$H(R) = \frac{1}{\beta}\ln E\left[Pe^{\beta R}\right], \tag{34}$$

where the parameter β is positive, and P is a random variable with $P > 0$ and $E(P) = 1$. Then

$$H'(R) = \frac{Pe^{\beta R}}{E\left[Pe^{\beta R}\right]}. \tag{35}$$

Furthermore, suppose that u_1 is the exponential utility function with parameter $\alpha > 0$. Using (32) we find that the optimal reimbursement is

$$R = -\frac{\alpha}{\alpha+\beta}X_1 - \frac{1}{\alpha+\beta}\ln P + d, \tag{36}$$

where d is an arbitrary constant. Note that this result is also of interest in the special case $P \equiv 1$, i.e., where H is the exponential principle of premium calculation of parameter β.

ACKNOWLEDGMENT

The exposition of this paper has benefitted from the helpful comments of a referee.

APPENDIX: QUADRATIC UTILITY FUNCTIONS

As in the case of exponential utility functions, some very explicit results can be obtained in the case of quadratic utility functions. Since quadratic utility functions can be criticized because of their increasing risk aversion, we present these results in an appendix.

It is assumed that the utility function of company i is

$$u_i(x) = x - \frac{x^2}{2s_i} \quad \text{for} \quad x \le s_i, \tag{10*}$$

where s_i is the level of saturation of company i. From (9) one finds that the Pareto optimal risk exchanges are of the form (7), where there is the following connection between quotas and side payments:

$$c_i = s_i - q_i s, \quad \text{with} \quad s = s_1 + \cdots + s_n. \tag{12*}$$

From (17) we find that

$$R_i = -X_i - f_i P + d_i, \text{ with } f_i = \frac{s_i - E(PX_i)}{1 + \mathrm{Var}(P)}, \tag{18*}$$

where d_i is an arbitrary constant. This and (14) leads to

$$P = 1 - \frac{X - \mu}{s - \mu}. \tag{19*}$$

Then the risk exchange corresponding to the equilibrium is given by (7) with

$$q_i = \frac{(s - \mu)(s_i - \mu_i) + \mathrm{Cov}(X, X_i)}{(s - \mu)^2 + \mathrm{Var}(X)}, \tag{22*}$$

and c_i as in (12*).

For the Bowley solution one finds that

$$P = 1 - b(X_1 - \mu_1),$$

with

$$b = \frac{\sqrt{(s_1 - \mu_1)^2 + \mathrm{Var}(X_1)} - (s_1 - \mu_1)}{\mathrm{Var}(X_1)}. \tag{28*}$$

REFERENCES

Bühlmann, H. (1970), *Mathematical Methods in Risk Theory.* New York: Springer.

Bühlmann, H. (1980), "An economic premium principle." *Astin Bulletin* **11**, 52–60.

Chan, F.-Y., and H. U. Gerber (1985), "The reinsurer's monopoly and the Bowley solution." *Astin Bulletin* **15**, 141–148.

Deprez, O., and H. U. Gerber (1985), "On convex principles of premium calculation." *Insurance: Mathematics and Economics* **4**, 179–189.

Gerber, H. U. (1985), "On additive principles of premium calculation." *Insurance: Mathematics and Economics* **4**, 249–251.

N. U. Prabhu [1]

A CLASS OF RUIN PROBLEMS

ABSTRACT

Von Bahr (1974) and others have investigated the ruin problem of insurance risk in the case where the total amount of claims received by the company is formulated as a compound renewal process. Such processes, under the name of renewal-reward processes, were introduced by Jewell (1967), who extended to them the fluctuation theory of random walks. In this paper we establish further results for compound renewal processes and apply them to Von Bahr's ruin problem and also to a generalized model in which the gross risk premium is represented by a subordinator with a drift, while the claim process is a compound renewal process.

1. INTRODUCTION AND PRELIMINARY RESULTS

Let $\{(\tau_k, X_k),\ k \geq 1\}$ be a sequence of independent random vectors with a common distribution on $[0,\infty) \times (-\infty,\infty)$, and let $T_0 = S_0 = 0$, $T_n = \tau_1 + \tau_2 + \cdots + \tau_n$, $S_n = X_1 + X_2 + \cdots + X_n$ $(n \geq 1)$. We shall ignore the trivial case where $P\{\tau_n = 0\} = 1$ $(n = 1)$. Then $\{(T_n, S_n),\ n \geq 0\}$ is a random walk on the right half-plane. Also let

$$N(t) = \max\{n :\ T_n \leq t\} \tag{1}$$

and

$$X(t) = \sum_{k=1}^{N(t)} X_k \tag{2}$$

(an empty sum being interpreted as zero). If the X_k are interpreted as the rewards earned at the renewal epochs T_k, then the total reward earned

[1] School of Operations Research and Industrial Engineering, College of Engineering, Cornell University, Ithaca, New York 14853

I. B. MacNeill and G. J. Umphrey (eds.), Actuarial Science, 63–78.

during a time-interval $(0,t]$ is given by $X(t)$. Jewell (1967), who introduced this process, called it the renewal-reward process and developed a fluctuation theory for it. If the τ_k are independent of the X_k and have a common density $\lambda e^{-\lambda t}$ $(0 < \lambda < \infty)$, then $N(t)$ becomes the simple Poisson process with parameter λ and $X(t)$ reduces to a compound Poisson process. In the general case it would therefore be appropriate to call $X(t)$ a compound renewal process. When $X_k \equiv 1$, $X(t)$ reduces to the renewal counting process. In general $X(t)$ is a special case of a semi-Markov process, in which the intervals between successive jumps are independent but do not follow the exponential distribution.

The fluctuation-theoretic behaviour of $X(t)$ obviously depends in a simple manner on that of the random walk $\{(T_n, S_n)\}$. However, there is no need to establish, *ab initio,* new results for this two-dimensional random walk (for example, the reflection principle), as Jewell did. This was observed by Von Bahr (1974) in his solution of the ruin problem, in which $X(t)$ represents the total amount of claim received by an insurance company during $(0,t]$; however, Von Bahr did not seem to have been aware of Jewell's work, and the present author did not know of Von Bahr's paper while working on the manuscript of his book (Prabhu, 1980, Chapter 4). In insurance risk theory one explanation of the possible dependence between τ_k and X_k is that the τ_k represent the intervals of time between the processing of successive claims X_k and may therefore depend on the size of the claims made. However, I am not aware of any numerical data demonstrating this phenomenon.

Von Bahr's main idea, expressed somewhat differently and without the constraints that he seems to have used, is the following. Since $\{T_n\}$ is a non-decreasing sequence, the first ladder epochs of the random walk $\{(T_n, S_n)\}$ are appropriately defined as

$$N = \min\{n : \ S_n > 0\}, \qquad \overline{N} = \min\{n : \ S_n \leq 0\}. \tag{3}$$

The corresponding ladder points are

$$(N, T_N, S_N) \quad \text{and} \quad (\overline{N}, T_{\overline{N}}, S_{\overline{N}}). \tag{4}$$

It suffices to consider only (T_N, S_N) and $(T_{\overline{N}}, S_{\overline{N}})$. Now for fixed $\theta > 0$, denote by

$$K_\theta\{I\} = E\left[e^{-\theta\tau_1}; X_1 \in I\right] \tag{5}$$

a (defective or discounted) probability measure. Its n-fold convolution is given by

$$K_{n\theta}\{I\} = E\left[e^{-\theta T_n}; S_n \in I\right] \quad (n \geq 1). \tag{6}$$

For the random walk $\{S_{n\theta},\ n \geq 0\}$ induced by K_θ, let (N_θ, S_{N_θ}) and $(\overline{N}_\theta, S_{\overline{N}_\theta})$ be the first ladder points. Then it is easily seen that

$$P\{N_\theta = n,\ S_{N_\theta} \in I\} = E\left[e^{-\theta T_n}; N = n,\ S_n \in I\right] \tag{7}$$

and

$$P\{\overline{N}_\theta = n,\ S_{\overline{N}_\theta} \in I\} = E\left[e^{-\theta T_n}; \overline{N} = n,\ S_{\overline{N}} \in I\right]. \tag{8}$$

From the theory of one-dimensional random walks, we immediately obtain the following results due to Jewell.

Theorem. For $\theta > 0$, ω real and $i = \sqrt{-1}$, we have

$$\begin{aligned}\chi(\theta,\omega) &= E\left(e^{-\theta T_N + i\omega S_N}\right)\\ &= 1 - \exp\left\{-\sum_1^\infty \frac{1}{n} E\left[e^{-\theta T_n + i\omega S_n};\ S_n > 0\right]\right\},\end{aligned} \tag{9}$$

$$\begin{aligned}\overline{\chi}(\theta,\omega) &= E\left(e^{-\theta T_{\overline{N}} + i\omega S_{\overline{N}}}\right)\\ &= 1 - \exp\left\{-\sum_1^\infty \frac{1}{n} E\left[e^{-\theta T_n + i\omega S_n};\ S_n \le 0\right]\right\},\end{aligned} \tag{10}$$

and

$$1 - E\left(e^{-\theta\tau_1 + i\omega X_1}\right) = [1 - \chi(\theta,\omega)][1 - \overline{\chi}(\theta,\omega)]. \tag{11}$$

The above theorem is central to Jewell's investigation of the fluctuation theory of $X(t)$. In Section 2, we shall apply the idea of a discounted probability measure to derive further results for $X(t)$. Our results reveal the remarkable similarities between the non-Markovian process $X(t)$ and processes with stationary independent increments (Lévy processes) in their fluctuation-theoretic behaviour.

In Section 3, we consider a compound renewal process with a negative drift, namely, $Y(t) = X(t) - \beta t$ ($\beta \ge 0$). We use the results of Section 2 to obtain the distribution of the supremum functional of Y and apply this result to derive an elegant solution to Von Bahr's ruin problem.

In the final section of the paper we consider a generalized insurance risk model in which the gross risk premium paid during a time-interval $(0, t]$ is represented by $X_1(t) + \beta t$, where X_1 is a subordinator and $\beta \ge 0$, while the claim process X_2 is a compound renewal process independent of X_1. The results of Section 2 are applied to this model to solve the ruin problem.

We now review some variants and extensions of the processes considered in this paper. A variant of the compound renewal process (2) occurs in Smith's (1955) investigation of cumulative processes, namely,

$$X(t) = \sum_{k=1}^{N(t)+1} X_k. \tag{12}$$

His main objective was to study the asymptotic behaviour of $X(t)$ and its moments. The processes discussed in Sections 3 and 4 of this paper are actually cumulative processes, namely, processes $Y(t)$ of bounded variation, which have an imbedded renewal process $\{T_n\}$ such that the random variables $Y(T_n) - Y(T_{n-1})$ $(n \geq 1)$ are mutually independent.

Suppose that the X_k are non-negative variables, and consider the process

$$Z(\tau) = \sum_{k=1}^{N(\tau)} X_k + \tau \qquad (\tau \geq 0). \tag{13}$$

Thus Z is a compound renewal process with a unit drift. The range of this process is given by

$$R = \bigcup_{n=0}^{\infty} [T_n + S_n,\ T_{n+1} + S_n). \tag{14}$$

Let

$$L(t) = |R \cap (0,t]|, \tag{15}$$

where $|A|$ is the Lebesgue measure of the set A. With a very different formulation, Takács (1957) gave an exhaustive treatment of the problem of finding possible limit distributions for $L(t)$. However, many of his tedious calculations can perhaps be avoided by observing that L is the right continuous inverse of Z (see Prabhu, 1986, Chapter I).

We mention two extensions of the basic ideas behind (2). In the first of these we consider the sequence $\{(\tau_k, X_k)\}$, where τ_k is the r-dimensional vector $(\tau_k^{(1)}, \tau_k^{(2)}, \ldots, \tau_k^{(r)})$ with $\tau_k^{(i)} \geq 0$ $(i = 1, 2, \ldots, r)$. For this we can obtain results similar to (9)–(11) with θt replaced by $\theta_1 t_1 + \theta_2 t_2 + \cdots + \theta_r t_r$. For $r = 2$ an application to single server queues has been described by Prabhu (1980, chapter 2).

As a second extension, consider the random walk $\{(T_n, S_n)\}$ on the plane, with $\nu_t = \min\{n : T_n > t\}$ and

$$X(t) = S_{\nu(t)}. \tag{16}$$

This process was investigated by Gut and Janson (1983), who also described several applications.

The following elementary result reveals the connection between the compound renewal process (2) and the compound Poisson.

Theorem 1. If the distribution of (τ_k, X_k) is defective, then the limit distribution of $X(t)$ as $t \to \infty$ is compound Poisson.

Proof. Suppose $P\{\tau_k < \infty,\ X_k < \infty\} = p < 1$. Then the renewal counting process $N(t) \to N$ almost surely (a.s.) as $t \to \infty$, where N is a random variable with the distribution $\{(1-p)p^n\ (n \geq 0)\}$. Therefore, as $t \to \infty$, $X(t) \to X$ (a.s.) where

$$X = \sum_{k=1}^{N} X_k. \tag{17}$$

Now, given $N = n$, the random variables $X_1, X_2, \ldots, X_N$ are mutually independent and have the proper distribution K_1, where

$$K_1\{I\} = P\{\tau_k < \infty,\ X_k \in I\}/p, \tag{18}$$

as can be easily verified. Therefore the distribution of (17) appears as the geometric mixture of convolutions, which is a compound Poisson (see Feller, 1971, Chapter XI).

In the remaining part of this paper, we shall assume that (τ_k, X_k) has a proper distribution, and ignore the case where $X_k = 0$ a.s.

2. FLUCTUATION THEORY OF COMPOUND RENEWAL PROCESSES

The fluctuation theory of the compound renewal process $X(t)$ is concerned with the limit behaviour of the functionals

$$M(t) = \sup_{0 \leq s \leq t} X(s), \qquad m(t) = \inf_{0 \leq s \leq t} X(s). \tag{19}$$

This behaviour follows directly from that of the corresponding functionals

$$M_n = \max(0, S_1, S_2, \ldots, S_n), \quad m_n = \min(0, S_1, S_2, \ldots, S_n) \tag{20}$$

of the random walk $\{S_n\}$. In fact we have

$$M(t) = M_{N(t)}, \qquad m(t) = m_{N(t)} \tag{21}$$

and of course $X(t) = S_{N(t)}$ by definition. To derive the results concerning (19) we denote by $\pi_n = \min\{m \leq n :\ S_m = M_n\}$ the first epoch at which the maximum M_n is achieved, and define

$$\begin{aligned} \zeta_+ &= \{n \geq 0 :\ M_n - S_n = 0, \quad \pi_n = n\}, \\ \zeta_- &= \{n \geq 0 :\ S_n - m_n = 0\}, \end{aligned} \tag{22}$$

the ladder set of $\{S_n\}$. Clearly,

$$\varsigma_+ = \{N_k,\ k \geq 0\}, \qquad \varsigma_- = \{\overline{N}_k,\ k \geq 0\}, \tag{23}$$

where the N_k are the successive strong ascending ladder epochs and $\overline{N}_k$ the weak descending ladder epochs of $\{S_n\}$. The random variables T_n do not play any role in these definitions; however, it should be noted that

$$X(T_{N_k}) = M_{N_k} \quad \text{and} \quad X(T_{\overline{N}_k}) = m_{\overline{N}_k}. \tag{24}$$

From the theory of one-dimensional random walks (see Prabhu, 1986, Chapter II) we obtain the following result which we state without proof.

Theorem 2. Suppose that the distribution of X_k is not concentrated either on $[0,\infty)$ or on $(-\infty,0]$. Then the compound renewal process $X(t)$ belongs to one of the following mutually exclusive types:

(i) $X(t)$ drifts to $+\infty$. Here

$$M(t) \to \infty, \quad m(t) \to m > -\infty, \quad X(t) \to \infty \text{ a.s.} \tag{25}$$

(ii) $X(t)$ drifts to $-\infty$. Here

$$M(t) \to m < \infty, \quad m(t) \to -\infty, \quad X(t) \to -\infty \text{ a.s.} \tag{26}$$

(iii) $X(t)$ oscillates. Here

$$-\infty = \liminf_{t\to\infty}\ X(t) < \limsup_{t\to\infty}\ X(t) = +\infty \text{ a.s.} \tag{27}$$

We wish to derive the distributions of $M(t)$ and $m(t)$. For this purpose let us denote the distribution of $X(t)$ by F_t, so that

$$F_t\{I\} = P\{X(t) \in I\} \tag{28}$$

for any interval I. For fixed $\theta > 0$, let us also denote

$$\mu_\theta\{I\} = \int_0^\infty e^{-\theta t} F_t\{I\} dt. \tag{29}$$

It is obvious that $\theta\mu_\theta$ is a probability measure on $(-\infty,\infty)$. It turns out that it is a compound Poisson measure, as shown by the following lemma.

Lemma 1. The measure $\theta\mu_\theta$ is compound Poisson, with the Lévy measure

$$\nu_\theta\{I\} = \sum_{n=1}^{\infty} \frac{1}{n} E\left[e^{-\theta T_n}; S_n \in I\right]. \tag{30}$$

Proof. We have

$$\begin{aligned}
F_t\{I\} &= \sum_{n=0}^{\infty} P\{N(t)=n,\ X(t)\in I\} \\
&= \sum_{n=0}^{\infty} \int_0^t \int_I P\{T_n\in d\tau,\ S_n\in dx,\ T_{n+1}>t\} \\
&= \sum_{n=0}^{\infty} \int_0^t \int_I P\{T_n\in d\tau,\ S_n\in dx\} P\{T_{n+1}>t|T_n=\tau, S_n=x\} \\
&= \sum_{n=0}^{\infty} \int_0^t P\{T_n\in d\tau,\ S_n\in I\}\{T_1>t-\tau\},
\end{aligned}$$

since $T_{n+1}-T_n$ is independent of (T_n, S_n) and has the same distribution as T_1. This yields the result

$$\theta\mu_\theta\{I\} = c_\theta \sum_{n=0}^{\infty} E\left[e^{-\theta T_n};\ S_n\in I\right], \tag{31}$$

where $c_\theta = 1-E(e^{-\theta\tau_1})$. To prove that $\theta\mu_\theta$ is compound Poisson, let us denote by

$$\phi(\theta,\omega) = E\left(e^{-\theta\tau_k+i\omega X_k}\right) \qquad (\theta>0,\ \omega \text{ real},\ i=\sqrt{-1}) \tag{32}$$

the characteristic function (c.f.) of (τ_k, X_k). Then (31) yields the result

$$\Phi_\theta(\omega) = \int_{-\infty}^{\infty} e^{i\omega x}\theta\mu_\theta\{dx\} = \frac{1-\phi(\theta,0)}{1-\phi(\theta,\omega)}. \tag{33}$$

We have

$$\begin{aligned}
\log\Phi_\theta(\omega) &= \sum_{n=1}^{\infty} \frac{1}{n}\left[\phi(\theta,\omega)^n - \phi(\theta,0)^n\right] \\
&= \sum_{n=1}^{\infty} \frac{1}{n} E\left[e^{-\theta T_n}\left(e^{i\omega S_n}-1\right)\right] \\
&= \int_{-\infty}^{\infty} (e^{i\omega x}-1)\nu_\theta\{dx\}
\end{aligned} \tag{34}$$

with ν_θ defined by (30). Since

$$\nu_\theta\{(-\infty,\infty)\} = \sum_{n=1}^{\infty}\frac{1}{n}E\left(e^{-\theta T_n}\right) = \sum_{n=1}^{\infty}\frac{1}{n}\phi(\theta,0)^n < \infty, \tag{35}$$

it follows that $\theta\mu_\theta$ is a compound Poisson measure.

Our objective is to establish a factorization result for the probability measure $\theta\mu_\theta$. For this purpose we introduce the positive Borel measures

$$\mu_\theta^+\{I\} = \sum_{n=0}^{\infty} E\left[e^{-\theta T_n};\ n \in \varsigma_+,\ S_n \in I\right] \tag{36}$$

and

$$\mu_\theta^-\{I\} = \sum_{n=0}^{\infty} E\left[e^{-\theta T_n};\ n \in \varsigma_-,\ S_n \in I\right] \tag{37}$$

for fixed $\theta > 0$. Clearly, the measures are concentrated on $[0,\infty)$ and $(-\infty, 0]$ respectively, and moreover, they are totally finite since

$$\begin{aligned}\mu_\theta^+\{[0,\infty)\} &= \sum_{n=0}^{\infty} E\left[e^{-\theta T_n};\ n \in \varsigma_+\right] \leq \sum_{n=0}^{\infty} E\left(e^{-\theta T_n}\right) \\ &= \sum_{n=0}^{\infty} \left[E\left(e^{-\theta \tau_1}\right)\right]^n < \infty \end{aligned} \tag{38}$$

and similarly $\mu_\theta^-\{(-\infty,0]\} < \infty$. We consider the normed measures

$$\hat{\mu}_\theta^+\{I\} = \mu_\theta^+\{I\}/\mu_\theta^+\{[0,\infty)\} \tag{39}$$

and

$$\hat{\mu}_\theta^-\{I\} = \mu_\theta^-\{I\}/\mu_\theta^-\{(-\infty,0]\}. \tag{40}$$

We have the following results.

Theorem 3. (Wiener-Hopf factorization for $X(t)$.) The probability measure $\theta\mu_\theta$ on $(-\infty,\infty)$ can be expressed in the form

$$\theta\mu_\theta = \hat{\mu}_\theta^+ * \hat{\mu}_\theta^- \qquad (\theta > 0) \tag{41}$$

where $\hat{\mu}_\theta^+$ and $\hat{\mu}_\theta^-$ are the probability measures (39) and (40), concentrated on $[0,\infty)$ and $(-\infty,0]$ respectively, and $*$ is the convolution operator. Moreover, among such measures this representation is unique except for translations.

Proof. For the one-dimensional random walk $\{S_n\}$, we have

$$P\{S_n \in I\} = \sum_{\ell=0}^{n} \int_{0-}^{\infty} P\{\ell \in \varsigma_+,\ S_\ell \in dx\} \cdot P\{n-\ell \in \varsigma_-,\ S_{n-\ell} \in I - x\}$$

(see Prabhu, 1986, Chapter III). Applying this result to the random walk $\{S_{n\theta}\}$ induced by the distribution (5), we obtain

$$E\left[e^{-\theta T_n};\ S_n \in I\right] = \sum_{\ell=0}^{n} \int_{0-}^{\infty} E\left[e^{-\theta T_\ell};\ \ell \in \varsigma_+,\ S_\ell \in dx\right] \cdot E\left[e^{-\theta T_{n-\ell}};\ n-\ell \in \varsigma_-,\ S_{n-\ell} \in I - x\right].$$

Summing this over $n = 0, 1, 2, \ldots$ we find that

$$\theta\mu_\theta = c_\theta \mu_\theta^+ * \mu_\theta^- \qquad (\theta > 0). \tag{42}$$

For $I = (-\infty, \infty)$ this gives

$$1 = c_\theta \mu_\theta^+\{[0,\infty)\} \cdot \mu_\theta^-\{(-\infty, 0]\}. \tag{43}$$

Dividing (42) by (43), we obtain the desired result (41). To complete the proof, we note that the probability measure $\theta\mu_\theta$ is compound Poisson by Lemma 1. Since

$$\hat{\mu}_\theta^+\{I\} = \sum_{k=0}^{\infty} E\left[e^{-\theta T_{N_k}};\ S_{N_k} \in I\right] \Big/ \sum_{k=0}^{\infty} E\left(e^{-\theta T_{N_k}}\right), \tag{44}$$

the proof of Lemma 1 also shows that $\hat{\mu}_\theta^+$ is a compound Poisson measure on $[0, \infty)$. Similarly $\hat{\mu}_\theta^-$ is a compound Poisson measure on $(-\infty, 0]$. Uniqueness of the factorization (41) follows from the fact that a compound Poisson measure on $(-\infty, \infty)$ can be represented in the desired form uniquely, except for translations.

Theorem 4. For $\theta > 0$, we have

$$\begin{aligned}\int_0^\infty \theta e^{-\theta t} P\{M(t) \in I,\ X(t) - M(t) \in J\} dt \\ = \int_0^\infty \theta e^{-\theta t} P\{X(t) - m(t) \in I,\ m(t) \in J\} dt \\ = \hat{\mu}_\theta^+\{I\} \hat{\mu}_\theta^-\{J\}\end{aligned} \tag{45}$$

for every pair of interval I, J.

Proof. Since $X(t)$ remains constant over the set $\cup_{k=0}^{\infty}[T_k, T_{k+1})$, the supremum and infimum of the process are achieved at the epochs T_{N_k} and $T_{\overline{N}_k}$

respectively, as was already observed in (24). We have

$$
\begin{aligned}
P\{M(t) \in & I,\ X(t)-M(t) \in J\} \\
&= \sum_{n=0}^{\infty} P\{N(t)=n,\ M(t)\in I,\ X(t)-M(t)\in J\} \\
&= \sum_{n=0}^{\infty}\sum_{m=0}^{n} \int_{\tau_1=0}^{t}\int_I P\{S_k < S_m\ (0\le k\le m-1), \\
&\qquad\qquad T_m \in d\tau_1,\ S_m \in dx\} \\
&\quad \cdot \int_{\tau_2=\tau_1}^{t} P\{S_k \le S_m\ (m\le k\le n),\ T_n\in d\tau_2, \\
&\qquad S_n - S_m \in J,\ T_{n+1} > t \mid T_m=\tau_1,\ S_m = x_1\} \\
&= \sum_{n=0}^{\infty}\sum_{m=0}^{n} \int_{\tau_1=0}^{t}\int_I P\{m\in \varsigma_+,\ T_m\in d\tau_1,\ S_m\in dx\} \\
&\quad \cdot \int_{\tau_2=\tau_1}^{t} P\{n-m\in\varsigma_-,\ T_{n-m}\in d\tau_2-\tau_1,\ S_{n-m}\in J\} \\
&\quad \cdot P\{T_1 > t-\tau_2\}.
\end{aligned}
$$

This leads to the result

$$\int_0^{\infty} \theta e^{-\theta t} P\{M(t)\in I,\ X(t)-M(t)\in J\}dt = c_\theta \mu_\theta^+\{I\}\mu_\theta^-\{J\}. \tag{46}$$

Taking $I=[0,\infty)$ and $J=(-\infty,0]$, we find that

$$1 = c_\theta\mu_\theta^+\{[0,\infty)\}\mu_\theta^-\{(-\infty,0]\}. \tag{47}$$

Dividing (46) by (47) we arrive at the result concerning $\{M(t),\ X(t)-M(t)\}$. The proof is completed in view of Lemma 2 below.

Lemma 2. We have

$$\{M(t),\ X(t)-M(t)\} \stackrel{d}{=} \{X(t)-M(t),\ m(t)\}. \tag{48}$$

Proof. The permutation

$$
\begin{aligned}
\{(\tau_1,X_1),\ &(\tau_2,X_2),\dots,(\tau_{n+1},X_{n+1})\} \to \\
&\{(\tau_n,X_n),\ (\tau_{n-1},X_{n-1}),\dots,(\tau_1,X_1),\ (\tau_{n+1},X_{n+1})\}
\end{aligned}
$$

leaves the distributions of M_n, m_n, S_n and $1_{N(t)=n}$ invariant. This results in the permutation of partial sums

$$\begin{aligned}&\{(T_r, S_r):\ 0 \leq r \leq n+1\}\\ &\rightarrow \{(T_n - T_{n-r},\ S_n - S_{n-r})(0 \leq r \leq n),\ (T_{n+1},\ S_{n+1})\}.\end{aligned}$$

From this we find that

$$P\{N(t) = n,\ M_n \in I,\ S_n - M_n \in J\} = P\{N(t) = n,\ S_n - m_n \in I,\ m_n \in J\}.$$

Adding this over $n = 0, 1, 2, \ldots$, we obtain the desired result since $X(t) = S_{N(t)}$, $M(t) = M_{N(t)}$, $m(t) = m_{N(t)}$.

Theorem 5. Let

$$A = \sum_{1}^{\infty} \frac{1}{n} P\{S_n \leq 0\}, \quad \text{and} \quad B = \sum_{1}^{\infty} \frac{1}{n} P\{S_n > 0\}. \tag{49}$$

Denote $\lim_{t\to\infty} M(t) = M \leq \infty$ and $\lim_{t\to\infty} m(t) = m \geq -\infty$. Then $M < \infty$ a.s. iff $B < \infty$, while $m > -\infty$ a.s. iff $A < \infty$, the distributions of M and m being given by

$$P\{M \in I\} = \hat{\mu}_0^+\{I\} \quad \text{and} \quad P\{m \in J\} = \hat{\mu}_0^-\{J\}, \tag{50}$$

where

$$\hat{\mu}_0^+\{I\} = \lim_{\theta\to 0+} \mu_\theta^+\{I\} \quad \text{and} \quad \hat{\mu}_0^-\{J\} = \lim_{\theta\to 0+} \hat{\mu}_\theta^-\{J\}. \tag{51}$$

Proof. From Theorem 4, we obtain

$$\int_0^\infty \theta e^{-\theta t} P\{M(t) \in I\} dt = \mu_\theta^+\{I\}.$$

Therefore by a Tauberian theorem

$$\begin{aligned}P\{M \in I\} &= \lim_{t\to\infty} P\{M(t) \in I\}\\ &= \lim_{\theta\to 0+} \int_0^\infty \theta e^{-\theta t} P[M(t) \in I\} dt = \hat{\mu}_0^+\{I\}.\end{aligned}$$

In the expression (44) for $\hat{\mu}_\theta^+\{I\}$, we note that

$$\lim_{\theta\to 0+} \sum_{k=0}^{\infty} E\left[e^{-\theta T_{N_k}};\ S_{N_k} \in I\right] = \sum_{k=0}^{\infty} P\{S_{N_k} \in I\} < \infty$$

since the last sum is a renewal measure. Also

$$\lim_{\theta\to 0+}\sum_{k=0}^{\infty} E\left(e^{-\theta T_{N_k}}\right) = \sum_{k=0}^{\infty} P\{T_{N_k} < \infty\}$$

and the second sum is finite if and only if there is a positive probability that ζ_+ is terminating, which happens if and only if $B < \infty$, as is known from the theory of one-dimensional random walks.

We conclude this section by some comments on the factorization (41). From Lemma 1 we know that the Lévy measure of $\theta\mu_\theta$ is ν_θ given by (30). The uniqueness of the factorization (41) implies that the Lévy measures of $\hat{\mu}_\theta^+$ and $\hat{\mu}_\theta^-$ are the restrictions of ν_θ to $[0,\infty)$ and $(-\infty,0]$ respectively. Denoting these latter measures by ν_θ^+ and ν_θ^-, we can therefore write

$$r_+(\theta,\omega) = \int_0^\infty e^{i\omega x}\hat{\mu}_\theta^+\{dx\} = \exp\left[\int_0^\infty (e^{i\omega x}-1)\nu_\theta^+\{dx\}\right] \tag{52}$$

and

$$r_-(\theta,\omega) = \int_{-\infty}^0 e^{i\omega x}\hat{\mu}_\theta^-\{dx\} = \exp\left[\int_{-\infty}^0 (e^{i\omega x}-1)\nu_\theta^-\{dx\}\right]. \tag{53}$$

The factorization (41) can then be written as

$$\int_0^\infty \theta e^{-\theta t} E[e^{i\omega X(t)}]dt = r_+(\theta,\omega) r_-(\theta,\omega). \tag{54}$$

In the case where $N(t)$ is a simple Poisson process independent of the X_k, we have already observed that $X(t)$ reduces to a compound Poisson process; we have

$$E[e^{i\omega X(t)}] = e^{-\lambda t[1-\psi(\omega)]},$$

where $\psi(\omega) = E(e^{i\omega X_1})$, and the identity (54) reduces to

$$\frac{\theta}{\theta+\lambda-\lambda\psi(\omega)} = r_+(\theta,\omega)r_-(\theta,\omega). \tag{55}$$

This is a special case of the Wiener-Hopf factorization for Lévy processes established by Rogozin (1966). The remarkable feature of (41) is that the factorization holds for the non-Markovian process $X(t)$.

3. THE RUIN PROBLEM FOR COMPOUND RENEWAL PROCESSES

In this section we consider the process

$$Y(t) = X(t) - \beta t, \tag{56}$$

where X is the compound renewal process described in Section 1 and β is a non-negative constant. We can view $Y(t)$ as the total reward obtained during $(0,t]$ minus a fee which is paid continuously at a constant rate β. (If the fee is paid only at the epochs T_n, then it merely changes the definition of X_k.) Our interest in this process is motivated by the insurance risk model (Von Bahr, 1974) in which the total amount of claims received by the company during a time-interval $(0,t]$ is represented by $X(t)$, and β is the gross risk premium. The risk reserve at time t is given by

$$Z(t) = x + \beta t - X(t), \tag{57}$$

where x (≥ 0) is the initial reserve. The ruin time is given by

$$T(x) = \inf\{t > 0 : \; Z(t) < 0\}. \tag{58}$$

We have

$$P\{T(x) > t\} = P\{M(t) \leq x\}, \tag{59}$$

where M is the supremum functional of the process (56). Our problem thus reduces to that of finding the distribution of M. Now an inspection of the sample function of Y shows that the supremum of Y coincides with that of the compound renewal process

$$X_1(t) = \sum_{k=1}^{N(t)} (X_k - \beta\tau_k), \tag{60}$$

since we can write

$$Y(t) = X_1(t) - \beta(t - T_{N(t)}). \tag{61}$$

From Theorem 4 and (52) we obtain the following result.

Theorem 6. Let $M(t)$ be the supremum functional of the process (56). Then, for $\theta > 0$, we have

$$\int_0^\infty \theta e^{-\theta t} E[e^{i\omega M(t)}]dt = \exp\left[\int_0^\infty (e^{i\omega x} - 1)\nu_\theta^+\{dx\}\right], \tag{62}$$

where

$$\nu_\theta^+\{I\} = \sum_{n=1}^{\infty} \frac{1}{n} E\left[e^{-\theta T_n};\ S_n - \beta T_n \in I \cap (0,\infty)\right]. \tag{63}$$

In view of (59), Theorem 6 provides the solution to the ruin problem, but for ease of reference, we state the following result which is essentially equivalent to the one given by Von Bahr (1974).

Theorem 7. For the ruin time (58) we have

$$1 + i\omega \int_0^\infty e^{i\omega x} E[e^{-\theta T(x)}]dx = \exp\left[\int_0^\infty (e^{i\omega x} - 1)\mu_\theta^+\{dx\}\right]$$
$$(\theta > 0, \quad \mathrm{Im}(\omega) > 0). \tag{64}$$

Proof. We have

$$i\omega\theta \int_0^\infty \int_0^\infty e^{i\omega x - \theta t} P\{T(x) > t\}dtdx = i\omega \int_0^\infty e^{i\omega x}[1 - Ee^{-\theta T(x)}]dx$$
$$= -1 - i\omega \int_0^\infty e^{i\omega x} E[e^{-\theta T(x)}]dx$$

and

$$i\omega\theta \int_0^\infty \int_0^\infty e^{i\omega x - \theta t} P\{M(t) \le x\}dxdt = -\theta \int_0^\infty e^{-\theta t} E[e^{i\omega M(t)}]dt.$$

In view of (59) these two results yield the desired result.

In the case where $X(t)$ is a compound Poisson process with parameter $\lambda(0 < \lambda < \infty)$, the Lévy measure $\nu_\theta^+\{I\}$ for $I \subset (0,\infty)$ is given by

$$\nu_\theta^+\{I\} = \sum_{i}^{\infty} \frac{1}{n} \int_0^\infty e^{-\theta t} e^{-\lambda t} \lambda^n \frac{t^{n-1}}{(n-1)!} P\{S_n \in I + \beta t\}$$
$$= \int_0^\infty t^{-1} e^{-\theta t} P\{X(t) \in I + \beta t,\ N(t) > 0\}dt$$
$$= \int_0^\infty t^{-1} e^{-\theta t} P\{Y(t) \in I\}dt \tag{65}$$

since $N(t) = 0$ implies $y(t) = -\beta t \le 0$. Substituting (65) into (64) we obtain the solution derived by Cramér (1955) by Wiener-Hopf techniques.

4. THE RUIN PROBLEM FOR A GENERALIZED MODEL

In the standard model for insurance risk, the gross risk premium is assumed to be paid at a constant rate β, as in Section 3. However, the variability in the amounts insured would imply that the company receives large as well as small amounts of premiums, with a large proportion of small premiums. An appropriate representation of the situation could therefore be provided by a Lévy process $X_1(t)$ with non-decreasing sample functions (subordinator) plus a drift βt ($\beta \geq 0$). We envisage $X_1(t)$ to be a process other than a compound Poisson on $[0, \infty)$, although the following result applies also to this case. The fact that $X_1(t)+\beta t \geq 0$ implies that the company is doing predominantly whole life insurance business. As in Section 3, we shall represent the total amount of claims received by the company during a time-interval $(0, t]$ by a compound renewal process, which we denote by $X_2(t)$. We shall assume that the processes X_1 and X_2 are independent. The risk reserve at time t is then given by

$$Z(t) = x + \beta t + X_1(t) - X_2(t). \tag{66}$$

For the ruin time (58), we once again have the relation (59), where M is the supremum functional of the process

$$Y(t) = X_2(t) - X_1(t) - \beta t. \tag{67}$$

Let $\{(\tau_k, X_k)\}$ be the sequence of random variables that generates $X_2(t)$ as in Section 1, and $Y_k = X_1(T_k) - X_1(T_{k-1})$ ($k \geq 1$). We can then write

$$Y(t) = X(t) - [X_1(t) - X_1(T_{N(t)})] - \beta(t - T_{N(t)}), \tag{68}$$

where

$$X(t) = \sum_{k=1}^{N(t)} (X_k - Y_k - \beta\tau_k). \tag{69}$$

As in Section 3, we need consider only the supremum functional of the compound renewal process $X(t)$. We thus obtain Theorem 7, with the measure ν_θ^+ in this case being given by

$$\nu_\theta^+\{I\} = \sum_1^\infty \frac{1}{n} E\left[e^{-\theta T_n};\ S_n - X_1(T_n) - \beta T_n \in I\right]$$

for $I \subset (0, \infty)$.

REFERENCES

Cramer, H. (1955), *Collective Risk Theory.* Jubilee Volume of the Skandia Insurance Company, Ab. Nordiska Bookhandeln, Stockholm.

Feller, W. (1971), *An Introduction to Probability Theory and Its Applications,* Volume 2, 3rd edition. New York: Wiley and Sons.

Gut, A., and S. Janson (1983), "The limiting behaviour of certain stopped sums and some applications". *Scandinavian Journal of Statistics* **10,** 281–292.

Jewell, W. S. (1967), "Fluctuations of a renewal-reward process". *Journal of Mathematical Analysis and Its Applications* **19,** 309–329.

Prabhu, N. U. (1980), *Stochastic Storage Processes.* New York: Springer-Verlag.

Prabhu, N. U. (1986), *Random Walks, Renewal Theory and Regenerative Phenomena.* To appear.

Rogozin, B. A. (1966), "On the distribution of functionals related to boundary problems for processes with independent increments". *Theory of Probability and Its Applications* **11,** 580–591.

Smith, W. L. (1955), "Regenerative stochastic processes". *Proceedings of the Royal Society,* Series A **232,** 6–31.

Takács, L. (1957), "On certain sojourn time problems in the theory of stochastic processes". *Acta Mathematica Academia Scientica Hungaricum* **8,** 169–191.

Von Bahr, B. (1974), "Ruin probabilities expressed in terms of ladder height distributions". *Scandinavian Actuarial Journal* 190–204.

S. David Promislow [1]

COMPARING RISKS

1. INTRODUCTION

In this paper we consider the problem of comparing probability distributions by their degree of risk. We survey one method of making such a comparison and discuss various actuarial applications. Generalizations and simplified proofs of some of the known mathematical results are given. A somewhat similar theme was considered by Goovaerts *et al.* (1982), but for the most part our applications and methods are different.

The original concepts appeared in the work of Hardy, Littlewood and Pólya. Consider two n-dimensional vectors $\mathbf{a} = (a_1, a_2, \ldots, a_n)$ and $\mathbf{b} = (b_1, b_2, \ldots, b_n)$ such that

$$\sum_{i=1}^{n} a_i = \sum_{i=1}^{n} b_i.$$

One says that $\mathbf{a}$ is *majorized* by $\mathbf{b}$, denoted by $\mathbf{a} \preceq \mathbf{b}$, if

$$\sum_{i=1}^{t} a_{[i]} \leq \sum_{i=1}^{t} b_{[i]} \text{ for } t = 1, 2, \ldots, n,$$

where $a_{[i]}$ denotes the ith largest element of the entries in $\mathbf{a}$. Hardy *et al.* (1929, 1952) proved the following result.

Theorem 1.1. The following are equivalent.

(a) $\mathbf{a} \preceq \mathbf{b}$.

(b) $\sum_{i=1}^{n} g(a_i) \leq \sum_{i=1}^{n} g(b_i)$ for all convex functions g.

(c) There is a doubly stochastic $n \times n$ matrix $\mathbf{P}$ such that $\mathbf{Pb} = \mathbf{a}$.

See Marshall and Olkin (1979) for an extensive discussion and generalizations. In this paper we consider the generalization that arises when we view

[1] Department of Mathematics, York University, Downsview, Ontario M3J 1P3

I. B. MacNeill and G. J. Umphrey (eds.), Actuarial Science, 79–93.

an n-dimensional vector $\mathbf{a}$ as representing the probability distribution which takes each of the values a_i with probability $1/n$. This leads to the following definition and theorem.

Definition. Define a partial order on the set of all probability distributions on the real line as follows. Given two such distributions F and G we will say that F is *less risky* than G and write $F \preceq G$ if

(a) The expectations exist and are equal. That is

$$\int_{-\infty}^{\infty} x dF(x) = \int_{-\infty}^{\infty} x dG(x) \tag{1.1}$$

and

(b)

$$\int_{t}^{\infty} (x-t) dF(x) \leq \int_{t}^{\infty} (x-t) dG(x) \tag{1.2}$$

for all real t.

If X and Y are random variables defined on the same probability space we will write $X \preceq Y$ to signify that $F \preceq G$, where X and Y have distributions F and G respectively.

The analogous result to Theorem 1.1 is given by the following Theorem.

Theorem 1.2. For distributions F and G satisfying (1.1), the following conditions are equivalent.

(a) $F \preceq G$.

(b) For all convex functions $g : \mathcal{R} \to \mathcal{R}$ ($\mathcal{R}$ denoting the set of real numbers)

$$\int_{-\infty}^{\infty} g(x) dF(x) \leq \int_{-\infty}^{\infty} g(x) dG(x)$$

provided that the integrals exist.

(c) There exist random variables X and Y defined on the same probability space, with distributions F and G respectively, such that $E(Y \mid X) = X$.

Remarks

(1) The connection between Part (c) in the two theorems is that the entry p_{ij} of the matrix P given in Theorem 1.1, equals the probability that Y takes the value of b_j given that X takes the value of a_i.

(2) The terminology of "less risky" is taken from Rothschild and Stiglitz (1970) and is justified by part (b) of the theorem which shows that all risk adverse individuals gain utility by choosing the less risky alternative.

(3) In view of (1.1), inequality (1.2) is equivalent to

$$\int_{-\infty}^{t} (t-x)dF(x) \le \int_{-\infty}^{t} (t-x)dG(x) \tag{1.3}$$

for all real t.

(4) There are many variations in definitions and in terminology. Some writers drop the equal mean assumption and define a partial order by (1.2) only. This is referred to as the *convex* ordering by Stoyan (1983), while it is called a *second order stop loss dominance* by Goovaerts *et al.* (1982). This, of course, reflects the fact that if the random variables represent total claims on a class of insurance policies, then it is immediate from (1.2) that all net stop-loss premiums are smaller for the less risky distribution. A analogous statement to Theorem 1.2 holds for this partial order. Part (b) holds for *nondecreasing* convex functions and in Part (c) the equality sign is replaced by "greater than or equal to".

One can also consider formula (1.3) by itself. The *dual* of such a partial order is referred to by Stoyan (1983) as the *concave* ordering and throughout the economic and finance literature as *second order stochastic dominance.*

We maintain the equal mean assumption throughout this paper, but all results admit suitable modifications for other cases.

Another variation is to use (1.2), but replace the integrand by $(x-t)^\alpha$, for some real number α greater than 1. The resulting partial order would be called *stop-loss dominance or order* $\alpha+1$ by analogy to the terminology used by Goovaerts *et al.* (1982), for the case of integer α. Similarly, the dual of the order defined by (1.3) with this adjustment is known as *stochastic dominance of order* $\alpha+1$.

Remarks on the Proof of Theorem 1.2. Various versions of this theorem have appeared frequently in the literature. Although many of the proofs involve some restrictions, such as differentiability of g, or boundedness of the random variables, it is true in the generality given. It is immediate that (b) implies (a), since we see that (1.2) (respectively (1.3)) correspond to statement (b) for the particular convex functions α_t (respectively β_t), defined by

$$\alpha_t(x) = \begin{cases} 0, & \text{for } x \le t \\ x-t, & \text{for } x > t, \end{cases}$$

$$\beta_t(x) = \begin{cases} t-x, & \text{for } x \le t, \\ 0, & \text{for } x > t, \ t \in \mathcal{R}. \end{cases}$$

The converse then follows by noting that (a) immediately implies (b) for all g which can be written as linear combinations of α_t and β_t with positive coefficients, that is for all piecewise linear convex functions. By suitable approximations, the conclusion follows for all convex g. See Stoyan (1983, Theorem 1.3.1) for an alternative proof. Also see Chong (1974) for further generalizations along these lines.

The proof that (c) implies (a) is often given by using the equivalence of (a) and (b) and invoking the conditional form of Jensen's inequality. A simple direct proof follows. We derive a random variable formulation of (1.2). Suppose that (c) holds for the probability space (Ω, Σ, P). For any real t, let

$$A = \{\omega \in \Omega : X(\omega) \geq t\}.$$

Then

$$\begin{aligned}\int_\Omega [X(\omega) - t]^+ dP(\omega) &= \int_A [X(\omega) - t] dP(\omega) = \int_A [Y(\omega) - t] dP(\omega) \\ &\leq \int_A [Y(\omega) - t]^+ dP(\omega) \leq \int_\Omega [Y(\omega) - t]^+ dP(\omega).\end{aligned}$$

The second equality above follows from the conditional expectation hypothesis.

The most difficult part is the proof that (a) implies (c). It can be obtained from the discrete case by applying suitable weak convergence results. This was carried out for the case of bounded random variables by Rothchild and Stiglitz (1970). This work appears to be the first to use the formulation of Part (c) as given. There is also an interesting interpretation which we will refer to in § 3. Whitt (1980) observed that the general result, with the boundedness condition removed, essentially follows from a result of Strassen (1965, Theorem 8).

Throughout the literature, the favorite method of proving that one distribution is less risky than another seems to be by using Part (b) of Theorem 1.2. In many cases it is easier and/or more intuitive to use Part (c). Such proofs also have the advantage that they can produce concrete and meaningful respresentations of the random variables X and Y. The existence of these random variables is obtained by nonconstructive methods in the general proof that (a) implies (c). We will illustrate this in connection with the distribution of aggegate claims in § 3.

A useful, easily applied, sufficient condition for riskiness is the so-called "cut condition". Given F and G satisfying (1.1), it is not hard to see that F is less risky than G, provided the graph of the distribution function of F begins below that of G and then crosses it in exactly one point. That is, there is a point c, such that $F(t) \leq G(t)$ for $t < c$ and $G(t) \leq F(t)$ for $t > c$. The corresponding random variable formulation, which we will use later, follows.

Theorem 1.3. Let X and Y be random variables defined on a probability space such that $E(X) = E(Y)$. Suppose there is a real number c such that

(a) $X(\omega) < c \Rightarrow Y(\omega) \le X(\omega)$

and

(b) $X(\omega) > c \Rightarrow Y(\omega) \ge X(\omega)$.

Then

$$X \preceq Y.$$

2. BASIC APPLICATIONS

(1) *Jensen's inequality.* It is easy to verify that for a random variable X which takes a constant value of c, X is less risky than Y, for any Y which has an expectation of c. Part (b) of Theorem 1.2 immediately yields Jensen's inequality, frequently used in actuarial (as well as in many other) applications.

(2) *Optimal forms of insurance.* Suppose that one wants to insure against a loss by paying some fixed premium which is *less* than the expected value of the loss, so that one can arrange for only partial coverage. A well-known result of Arrow (1963) states that all risk averse individuals will maximize utility by choosing deductible insurance. See Bowers *et al.* (1982, § 1.5) for complete details. We can use Theorem 1.3 to simplify the derivation of this result. Let a denote the constant initial wealth of the individual, P the premium, L the loss, I the amount of insurance paid. The random variable of interest is then the resulting wealth after insuring,

$$W = a - P - (L - I).$$

We assume that $E(I) = P$ and that the value of I is always nonnegative. Let X denote W when the insurance pays all losses above a deductible of d. Then

$$L - I = \begin{cases} L & \text{if } L \le d \\ d & \text{if } L > d. \end{cases}$$

Let Y denote W for any other admissible form of insurance. We apply Theorem 1.3 with $c = a - P - d$. Part (a) holds vacuously since X is never less than c. Suppose that X is greater than c. We then have that

$$Y = a - P - (L - I) \ge a - P - L = X$$

showing that part (b) is satisfied. We conclude that the deductible form is less risky than any other.

(3) *The adjustment coefficient.* We will follow the notation used by Bowers *et al.* (1982, § 12.3). For simplicity we confine discussion to the continuous case, but the result given is true as well for the discrete adjustment coefficient described by Bowers *et al.* (1982, § 12.4). Suppose we have a risk process based on a nonnegative claim distribution X, a Poisson distribution for the claim frequencies, and premium loading θ. Define a function

$$h_X(r) = M_X(r) - 1 - (1+\theta)rE(X)$$

on the interval $(0, \gamma_X)$ where $(-\infty, \gamma_X)$ is the domain of $M_X(r)$ (the moment generating function of X). Then h has a unique zero R which is known as the adjustment coefficient. Moreover, $h(r)$ is negative for r less than R. The standard use of the adjustment coefficient is in probability of ruin formulas, but in its own right it can be used as a measure of risk, with higher values indicating less risk, as suggested by Bowers *et al.* (1982, Theorem 13.1, and Examples 12.3, 13.9, 13.10). For fixed θ, consider two claim distributions, X and Y, and let R_X and R_Y denote the respective adjustment coefficients. Then

$$X \preceq Y \text{ implies that } R_X \geq R_Y.$$

This result was given by Goovaerts and De Vylder (1983) but without a complete proof. A simple derivation can be obtained by using the convexity of the function $g(x) = e^{rx}$ and Theorem 1.2. Then $X \preceq Y$ implies that $M_X(r) \leq M_Y(r)$ and therefore that $h_X(r) \leq h_Y(r)$ for all r in the interval $(0, \gamma_Y)$. This shows that $\gamma_X \geq \gamma_Y$ and

$$h_X(R_Y) \leq h_Y(R_Y) = 0.$$

The conclusion follows from the above remarks concerning the form of the function h.

The same argument shows that for fixed X the adjustment coefficient decreases as θ increases.

It is clear that we do not need $X \preceq Y$ in the above derivation but only the weaker condition that $M_X(r) \leq M_Y(r)$ for all y in $(0, \gamma_Y)$. By a reasoning similar to that of Stoyan (1983, Formula 1.8.4b), we see that we also have $R_X \geq R_Y$ when X is "dominated by" Y under stop loss dominance of order α for all $\alpha \geq 1$. This was indicated for integer values of α by Goovaerts and De Vylder (1983).

(4) *Survival distributions.* Many of the applications of the type of partial order relations we are discussing have been in the field of reliability theory. Somewhat similar considerations apply to survival distributions. We will consider one simple example. Consider two survival distributions T and T', each representing the time until death of a life age x. What does it mean to

say that $T \preceq T'$? Integrating by parts, the inequalities (1.2) and (1.3) yield (letting primed symbols refer to T') that for all $s > 0$

$$\int_s^\infty {}_tp_x dt \leq \int_s^\infty {}_{t'}p_x dt$$

and

$$\int_0^s {}_tp_x dt \geq \int_0^s {}_{t'}p_x dt.$$

In other words, under T, the expected lifetime *after* any point of time is less than or equal to what it would be under T', and the reverse is true for the expected lifetime *before* any point of time. Classical actuarial methods would consider T and T' to be equivalent for insurance and annuity purposes since they have the same expectation. Even if one goes a step further and takes variance into account, it is conceivable that T and T' could have equal variances. However, (assuming $T \neq T'$) it is clear that distribution T' carries more risk, and the actuary might well wish to reflect this in ratemaking. For example, in the case of an insurance policy issued to a class of individuals with distribution T' as opposed to T, one can expect more claims in the early durations and fewer claims in the later durations.

This example illustrates the basic idea that one must sometimes look at more than just the mean and variance when comparing risks.

(5) *Normal distributions.* This is, of course, not primarily concerned with actuarial matters, but Example (4) above suggests that we clarify the role of the traditional measure of risk, namely the variance. Due to the convexity of x^2 the variance can never be greater for a less risky distribution. For normal distributions, a converse holds in the sense that $\sigma_1^2 \leq \sigma_2^2$ implies that $N(\mu, \sigma_1^2) \preceq N(\mu, \sigma_2^2)$. An easy proof of this fact is to use Part (c) of Theorem 1.2. Let "$=_d$" denote equality in distribution and consider random variables, $X =_d N(\mu, \sigma_1^2)$ and $Z =_d N(0, \sigma_2^2 - \sigma_1^2)$, with X and Z independent. Then clearly $E(X + Z \mid X) = X$ and $X + Z =_d N(\mu, \sigma_2^2)$.

3. THE DISTRIBUTION OF AGGREGATE CLAIMS

We now wish to consider in somewhat greater detail, a particular application which is of major importance in actuarial work. Let N be a nonnegative integer valued distribution, representing claim frequencies during some fixed period, and let X_i denote the amount of the ith claim. We are interested in the aggregate claims given by

$$S = X_1 + X_2 + \cdots + X_N.$$

We will consider different distributions for N, but the sequence $\{X_i\}$ will be the same distribution in all cases, and we will denote dependence on the claim frequency distribution by writing S as S_N.

A basic question then is the following: If $N \preceq M$, is $S_N \preceq S_M$?

The answer, of course, depends on the assumptions we make. The easiest case is when the $\{X_i\}$ are i.i.d. and independent of N. Even with these assumptions the answer to our question is no! Let X_i $(i = 1, 2)$ take the values 1 and -1 each with probability 1/2. Let N have a constant value of 1 and let M take the values 0 and 2 each with probability 1/2. Then S_N is distributed as X_i and S_M takes the values -2 with probability 1/8, 2 with probability 1/8 and 0 with probability 3/4. It is not hard to deduce that S_N and S_M are incomparable. For example, use the vector formulation as given in Theorem 1.1, with the vectors

$$\mathbf{a} = (-1,-1,-1,-1,1,1,1,1) \text{ and } \mathbf{b} = (-2,0,0,0,0,0,0,2).$$

This occurs because of the negative values in the claim size distribution. The extra risk in claim frequency is offset by the greater possibility of cancellation. Consider the basic question again under the more realistic assumption of nonnegative claims. The result is now true under the assumptions given above. A proof is given by Goovaerts *et al.* (1982, p. 141), based on a somewhat complicated convex inequality which the authors use to generalize the result to the nth order case (*ibid*, Theorem 5.14). Another type of generalization, which weakens the assumption that the $\{X_i\}$ are identically distributed is given by Stoyan (1983).

Theorem 3.1. (Stoyan, 1983, Prop. 2.2.5) Let $\{X_i\}$ be a sequence of nonnegative independent random variables such that

$$X_i \preceq X_{i+1}, \qquad i = 1, 2, \ldots,$$

and let N and M be nonnegative integer valued random variables independent of $\{X_i\}$. Then

$$M \preceq N \text{ implies that } S_N \preceq S_M.$$

The proof given by Stoyan (1983), attributable to A. A. Borokov, also uses the convex function formulation. As indicated in the introduction we wish to give a more intuitive derivation by invoking Part (c) of Theorem 1.2. We first recall some basic and well-known facts about our partial ordering.

Lemma 3.2. Suppose we are given two sequences of distributions, F_i, G_i, and a sequence α_i of positive numbers with $\sum \alpha_i = 1$, for $i = 0, 1, 2, \ldots$. Let F and G denote mixtures of F_i and G_i respectively with weights α_i.

$$\text{If } F_i \preceq G_i \text{ for each } i, \text{ then } F \preceq G.$$

Proof. This follows immediately, using either parts (b) or (c) of Theorem 1.2.

Lemma 3.3. Given distributions F, G, H,

$$F \preceq G \text{ implies } F * H \preceq G * H,$$

where * denotes convolution.

Proof. This is true for all partial orders generated by a class of functions invariant under translation, which clearly holds for convex functions. See Stoyan (1983, Definition 1.1.2 and Proposition 1.1.2).

The following result is of independent interest. Its proof is somewhat technical and will be given in § 4.

Theorem 3.4. Let N be a distribution on Z^+ (the nonnegative integers) with $E(N) = m$. Then N can be written as a mixture of *two-valued* distributions on Z^+ *each* with expectation m.

Rothschild and Stiglitz (1970) observed that Part (c) of Theorem 1.2 allows us to think of the relation of riskiness in the following way. If X is less risky than Y, and we view X as winnings in a lottery, then Y may be viewed as the winnings in a lottery which has *exactly* the same payoffs, except that instead of being paid in cash, the amounts are in the form of some other lottery ticket with the original winnings as the expectation. It is instructive to visualize Theorem 3.1 from this point of view. We will use the language of lotteries instead of claims and consider the easiest nontrivial case. We wish to compare the purchase of a single lottery ticket, with an alternative which gives you either two tickets or none with equal chance. According to Theorem 3.1, the second choice is more risky and the problem is to see this in terms of the Rothschild-Stiglitz formulation. One way is as follows. Imagine a second lottery, where instead of receiving your winnings in cash you get *another* ticket, plus a real number chosen uniformly from the unit interval. You then collect on both tickets if your real number is less than the ratio of the original winnings to the total winnings on both tickets. If your real number is greater than this ratio, you receive nothing. If the amount won on the first ticket is x and that on the second is y, then you win $(x + y)$ with probability $x/(x + y)$, so it is clear that your expectation is the same as the amount won on the first ticket. Moreover, by symmetry we see that the second lottery does indeed give you a probability of 1/2 of collecting on both tickets.

We can use the idea in this example to produce a general proof.

Proof of Theorem 3.1. By Theorem 1.2 (c) we can find a joint distribution such that $E(M \mid N) = N$. Now using Lemma 3.2, with the weights $\alpha_i = P(N = i)$, we can reduce to the case where N is constant. Then using Lemma 3.2 again, along with Theorem 3.4, we can reduce to the case that N is constant and M is two valued. Note that up to now this is perfectly general and requires no assumptions whatsoever on $\{X_i\}$. Using the independence assumption and Lemma 3.3, we can further assume that one of the values which M takes is zero, since we pass from there to the general case by adding an independent summand to each side of the inequality. We are thereby reduced to considering the following. Given any integers $0 < m < k$, let $S = X_1 + X_2 + \cdots + X_m$, $V = X_1 + X_2 + \cdots + X_k$, and let T take the value V with probability m/k, and 0 with probability $(k-m)/k$. We want to show that $S \preceq T$. The proof is by induction on m. Assume then that the conclusion is true when the index in S is less than or equal to m. Hence, $X_2 + X_3 + \cdots + X_m$ is less risky than a mixture of 0 with probability $(k-m)/(k-1)$ and $X_2 + X_3 + \cdots + X_k$ with a probability of $(m-1)/(k-1)$. Adding the independent summand X_1 and then invoking Lemma 3.2 we see that S is less risky than a mixture of X_1 with probability $(k-m)/(k-1)$ and V with probability $(m-1)/(k-1)$. Now again by the induction hypothesis, X_1 is less risky than a mixture of 0 with probability $1/k$ and V with probability $(k-1)/k$. The conclusion that $S \preceq T$ follows from Lemma 3.2.

It remains to derive the result for m equal to 1. It is sufficient to do this in the case where the X_i's are identically distributed, for $i = 1, 2, \ldots, k$. The general case follows by repeated applications of Lemma 3.3. Suppose then that each X_i is distributed as a random variable X defined on a probability space Ω and let Ω^* be the product space $\Omega^k \times I$ (where I denotes the unit interval. We define random variables S^* and T^* on Ω^* as follows. First, let r denote the ratio $\omega_1/(\omega_1 + \omega_2 + \cdots + \omega_k)$ where we take r to be 1 if the denominator takes the value of 0.

Now define

$$S^*(\omega_1, \omega_2, \ldots, \omega_k, t) = X(\omega_1),$$

$$T^*(\omega_1, \omega_2, \ldots, \omega_k, t) = \begin{cases} X(\omega_1) + X(\omega_2) + \cdots + X(\omega_k) & \text{if } t \le r \\ 0 & \text{if } t > r. \end{cases}$$

It is straightforward to verify that S^* is distributed as S, and $E(T^* \mid S^*) = S^*$. Using the fact that the X_i's are identically distributed we can also verify from a symmetry argument that T^* is distributed as T completing the proof.

The case where $m = 1$ in the proof of Theorem 3.3 can be generalized as follows.

Theorem 3.5. Let (Ω, Σ, P) be a probability space. Let G be a finite group acting as measure preserving transformations on Ω. Let X be any nonnegative random variable defined on Ω such that $E(X)$ exists, and let

$$Z = \sum_{\alpha \in G} X_\alpha \quad (\text{where } X_\alpha(\omega) = X[\alpha(\omega)]).$$

Let Y be a mixture of 0 with probability $(k-1)/k$ and Z with probability $1/k$ (where k is the cardinality of G). Then

$$X \preceq Y.$$

Proof. Define a new measure $\hat{P}$ on Ω by

$$\hat{P}(A) = \int_A [X(\omega)/Z(\omega)] dP(\omega),$$

where the integrand is taken equal to 1 if $Z(\omega) = 0$. In general $\hat{P}$ is *not* a probability measure.

Using the measure preserving property we have that for any α in G,

$$\hat{P}(A) = \int_{\alpha^{-1}(A)} [X_\alpha(\omega)/Z_\alpha(\omega)] dP(\omega).$$

Now Z_α is clearly equal to Z, and summing the above over all elements of G, it follows that for any G-invariant set A,

$$k\hat{P}(A) = \int_A [Z(\omega)/Z(\omega)] dP\omega)$$

and hence

$$\hat{P}(A) = (1/k)P(A) \text{ for invariant } A.$$

We now let (Ω^*, Σ^*) be the product measureable space $\Omega \times \{1,2\}$ and define a probability measure P^* on this space by

$$P^*(A,1) = \hat{P}(A),$$
$$P^*(A,2) = P(A) - \hat{P}(A).$$

Define random variables X^* and Y^* on $(\Omega^*, \Sigma^*, P^*)$ by

$$X^*(\omega, i) = X(\omega), \qquad i = 1,2,$$
$$Y^*(\omega, 1) = Z(\omega),$$
$$Y^*(\omega, 2) = 0.$$

It is clear that X^* is distributed as X. Any set B^* in the σ-algebra generated by X^* is of the form $(A,1) \cup (A,2)$, where A is in the σ-algebra generated by X, and therefore

$$\int_{B^*} Y^* dP^* = \int_A Z(\omega) d\hat{P}(\omega) = \int_A X(\omega) dP(\omega) = \int_{B^*} X^*(\omega) dP^*(\omega).$$

Hence $E(Y^* \mid X^*) = X^*$, which completes the proof.

We now show how this group formulation covers the case discussed in Theorem 3.5. Suppose we begin with a nonnegative random variable X_0 defined on a probability space Ω_0. Let Ω denote the product space $(\Omega_0)^k$. Define a random variable X on Ω by

$$X(\omega_1, \omega_2, \ldots, \omega_k) = X_0(\omega_1)$$

and a measure preserving transformation γ by

$$\gamma(\omega_1, \omega_2, \ldots, \omega_k) = (\omega_2, \omega_3, \ldots, \omega_k, \omega_1).$$

Let G be the cyclic group generated by γ. We now have the situation of Theorem 3.5, and the random variable Z of that theorem is just the sum of k independent copies of X.

4. PROOF OF THEOREM 3.4

Let m be any nonnegative real number which will remain fixed throughout the discussion. Let $D = \{\mathbf{k} = (k_1, k_2) : k_1$ and k_2 are nonnegative integers with $k_1 \leq m \leq k_2\}$. For $\mathbf{k}$ in D let $S(\mathbf{k})$ denote the unique distribution on Z^+ which takes the values k_1 and k_2 and has expectation m. That is (for $k_1 < k_2$),

$$P[S(\mathbf{k}) = k_1] = \frac{k_2 - m}{k_2 - k_1}, \quad P[S(\mathbf{k}) = k_2] = \frac{m - k_1}{k_2 - k_1}. \tag{4.1}$$

Now given any distribution N on Z^+ with $E(N) = m$, we will prove the theorem by finding for each $\mathbf{k}$ in D, a nonnegative constant $\alpha(\mathbf{k})$ such that $\sum_{\mathbf{k} \in D} \alpha(\mathbf{k}) = 1$ and for each $n \in Z^+$,

$$P[N = n] = \sum_{\mathbf{k} \in \mathbf{D}} \alpha(\mathbf{k}) P[S(\mathbf{k}) = n]. \tag{4.2}$$

Let $m = m_0 + f$ where $0 \leq f < 1$. The proof is by induction on m_0. Let $p(n) = P[N = n]$ and let

$$\sigma = \sum_{i=1}^{\infty} (i - f) p(m_0 + i).$$

Then

$$
\begin{aligned}
m = E(N) &= \sum_{i=1}^{\infty} ip(i) \\
&= p(1) + 2p(2) + \cdots + m_0 p(m_0) + m \sum_{i=1}^{\infty} p(m_0 + i) + \sigma \\
&\leq m(1 - p(0)) + \sigma, \qquad (4.3)
\end{aligned}
$$

showing that

$$\sigma \geq p(0)m. \qquad (4.4)$$

Case 1. $\sigma = p(0)m$.

Consider first the case where $f \neq 0$. Equality holds in (4.3) and we must have that

$$\sum_{i=1}^{\infty} p(m_0 + i) = 1 - p(0). \qquad (4.5)$$

For $i = 1, 2, \ldots$, let

$$\beta_i = (m_0 + i)p(m_0 + i)m^{-1}.$$

Now

$$m^{-1}\left[\sum_{i=1}^{\infty} \beta_i - \sigma\right] = \sum_{i=1}^{\infty} p(m_0 + i).$$

Using (4.5) and substituting for σ yields

$$\sum_{i=1}^{\infty} \beta_i = 1.$$

The desired conclusion is satisfied by taking $\alpha(\mathbf{k}) = \beta_i$ for $\mathbf{k} = (0, m_0 + i)$, and $\alpha(\mathbf{k}) = 0$ for any other pair of nonnegative integers, since for $\mathbf{k} = (0, m_0 + i)$, we see from (4.1) that

$$\beta_i P[S(\mathbf{k}) = m_0 + i] = \beta_i m/(m_0 + i) = p(m_0 + i).$$

If $f = 0$, then the equality in (4.3) shows that $p(n) = 0$ for $1 \leq i \leq m$ and (4.5) will again hold except with a lower limit of 0. The conclusion is the same as before except that the distribution concentrated at the integer m, i.e., $S(\mathbf{k})$ with $\mathbf{k} = (m, m)$, must be included with a weight of $p(m)$.

Note that if $m_0 = 0$, then Case 1 always holds, which starts the induction.

Case 2. $\sigma > p(0)m$.

Let

$$\sigma_t = \sum_{i=1}^{t}(i-f)p(m_0+i)$$

and choose the smallest t for which $\sigma_t \geq p(0)m$. Let $\sigma_0 = 0$. Then

$$\sigma_{t-1} \leq p(0)m < \sigma_{t-1} + (t-f)p(m_0+t). \tag{4.7}$$

Let

$$\gamma_i = (m_0+i)p(m_0+i)m^{-1}, \quad i = 1,2,\ldots,t-1,$$

and

$$\gamma_t = \left[p(0) - \frac{\sigma_{t-1}}{m}\right]\left[\frac{m_0+t}{t-f}\right], \tag{4.8}$$

so that

$$\sum_{i=1}^{t-1}\gamma_i(i-f)/(m_0+i) = \sigma_{t-1}/m \tag{4.9}$$

and from (4.8)

$$\sum_{i=1}^{t}\gamma_i(i-f)/(m_0+i) = p(0). \tag{4.10}$$

Multiplying the second inequality in (4.7) by $(m_0+t)/[m(t-f)]$ and using (4.8) we see that

$$\gamma_t \leq p(m_0+t)(m_0+t)m^{-1}.$$

Now letting

$$\gamma = \sum_{i=1}^{t}\gamma_i,$$

we have that

$$\gamma \leq \sum_{i=1}^{t}p(m_0+i)(m_0+i)m^{-1} \leq E(N)m^{-1} = 1.$$

The above discussion shows that there is a distribution Q on Z^+ such that $E(Q) = m$ and N can be written as a mixture of $S(0, m_0+1)$, $S(0, m_0+2), \ldots, S(0, m_0+t)$ and Q with weights $\gamma_1, \gamma_2, \ldots, \gamma_t, 1-\gamma$. Now (4.10) and (4.1) show that $P[Q=0] = 0$, so that $Q-1$ is a distribution on Z^+ with expectation $m-1$. By the induction hypothesis, we can write $Q-1$ as a mixture of two valued distributions each with expectation $m-1$. We can, therefore, write $Q = 1 + (Q-1)$ as a mixture of $S(\mathbf{k})$'s. This completes the proof.

REFERENCES

Arrow,K. J. (1963), "Uncertainty and the welfare economics of medical care." *American Economic Review* **53**, 941–973.

Bowers, N. L., H. U. Gerber, J. C. Hickman, D. A. Jones, and C. J. Nesbitt (1982), *Actuarial Mathematics*, Society of Actuaries.

Chong, K. (1974), "Some extensions of a theorem of Hardy, Littlewood, and Pólya and their applications." *Canadian Journal of Mathematics* **26**, 1321–1340.

Goovaerts, M. J., F. De Vylder, and J. Haezendonck (1982), "Ordering of risks: a review." *Insurance: Mathematics and Economics* **1**, 131–161.

Goovaerts, M. J., and F. De Vylder (1983), "Upper and lower bounds on infinite time ruin probabilities in case of constraint on claim size distributions." *Journal of Econometrics* **23**, 77–90.

Hardy, G. H., J. E. Littlewood, and G. Pólya (1929), "Some simple inequalities satisfied by convex functions." *Messenger of Mathematics* **58**, 145–152.

Hardy, G. H., J. E. Littlewood, and G. Pólya (1952), *Inequalities*. London: Cambridge University Press.

Marshall, A. W., and I. Olkin (1979), *Inequalities; Theory of Majorization and its Applications*. New York: Academic Press.

Rothschild, M., and J. E. Stiglitz (1970), "Increasing risk, 1: A definition." *Journal of Economic Theory* **2**, 225–243.

Stoyan, D. (1983), *Comparison Methods for Queues and Other Stochastic Models*. New York: Wiley and Sons.

Strassen, V. (1965), "The existence of probability measures with given marginals." *Annals of Mathematical Statistics* **36**, 423–429.

Whitt, W. (1980), "The effect of variability in the G1/G/s queue." *Journal of Applied Probability* **17**, 1062–1071.

Stuart Klugman [1]

INFERENCE IN THE HIERARCHICAL CREDIBILITY MODEL

1. INTRODUCTION

The problem of credibility estimation can be approached from a number of different directions (Gerber, 1982), but in most cases the authors stop with the determination of point estimators. Yet an investigator would also want some information about the error structure of these estimators. Variances of the estimators in an empirical Bayes setting for the basic model (Bühlmann and Straub, 1972) when the model variance is known are developed by Morris (1983a,b). Similar results with all variances unknown are developed by Klugman (1985b). Distributions for use in hypothesis tests are given in most linear models texts (e.g., Graybill, 1961). They are summarized for credibility models by Klugman (1985a).

In this paper inferences are obtained for the hierarchical model. This model was selected because it includes enough complexity to illustrate the effort that is required to construct these intervals and tests. In particular, the tests become important when deciding precisely which models should be used. Also, the basic model is a special case of this model.

The hierarchical model is based on having the various classes of insureds being partitioned into several groups. An example of this is worker's compensation insurance where the classes are occupations. It is reasonable to group these classes by industry classification. The randomness in the model allows for the overlapping of the class means since the grouping cannot be expected to be perfect.

In Section 2, the model will be presented in two forms and point estimators for the group and class means are derived. In Section 3, variance estimates for these estimators are presented. Finally, in Section 4, appropriate hypothesis tests are obtained.

---

[1] Department of Statistics and Actuarial Science, The University of Iowa, Iowa City, Iowa 52242

I. B. MacNeill and G. J. Umphrey (eds.), Actuarial Science, 95–113.

2. THE MODEL AND POINT ESTIMATORS

The hierarchical model is constructed by letting the observations be Y_{ijt}, $i = 1,\ldots,k$, $j = 1,\ldots m_i$, and $t = 1,\ldots n_{ij}$. Let $N = \sum_{ij} n_{ij}$ be the total number of observations and $M = \sum_i m_i$ be the number of classes. Given structural parameters $\mu, A, U_1,\ldots,U_k$, and V, the distribution of Y_{ijt} can be expressed either of two ways:

(1) "Bayesian"

$$\begin{aligned} Y_{ijt} \mid \theta_{ij} &\sim N(\theta_{ij}, V/p_{ijt}), \\ \theta_{ij} \mid \beta_i &\sim N(\beta_i, U_i), \\ \beta_i \mid \mu &\sim N(\mu, A); \end{aligned}$$

(2) "Variance Components"

$$Y_{ijt} = \mu + a_i + b_{ij} + e_{ijt},$$

where

$$\begin{aligned} a_i &\sim N(0, A), \\ b_{ij} &\sim N(0, U_i), \\ e_{ijt} &\sim N(0, V/P_{ijt}). \end{aligned}$$

In both settings all the random variables are assumed to be mutually independent. The relationships connecting the models are:

$$\begin{aligned} \theta_{ij} &= \mu + a_i + b_{ij}; \\ \beta_i &= \mu + a_i. \end{aligned}$$

It is assumed that each Y_{ijt} is an average based on P_{ijt} (or at least a number proportional to P_{ijt}) observations where i is the group, j the class, and t the observation (usually a year) in class ij.

Model (1) is a hierarchical extension of those used in most credibility work (e.g., Bühlmann, 1967; Morris, 1983a). The Bayesian framework comes from viewing the distributions as priors and hyperpriors. Such an analysis requires the structural parameters to be specified in advance. This model was discussed in detail by Lindley and Smith (1972). Most practitioners use a form of empirical Bayes analysis, where method of moments estimates replace the structural parameters in the posterior analysis. This is advocated by Bühlmann and Straub (1972) and by ISO (1980). A more formal empirical Bayes approach using methods of moments estimators and vague priors is presented by Morris (1983a,b). This is the approach that will be used in this paper when creating confidence intervals. In many of the papers cited above, the normal distribution assumption is not used. Instead, estimators are restricted to being linear functions of the observations, yielding identical minimum mean squared error estimators. This will not be sufficient for inference purposes as more than the first two moments of the three distributions will be required.

Model (2) is a standard linear model representation of the observations. It produces exactly the same covariance structure as model (1) but presents it without reference to higher levels of random variables. This formulation provides access to the estimation theory for variance component models (e.g., Graybill, 1961). Most such analyses are concerned with estimation of the variances and not with the estimation of the realizations of the random variables a_i and b_{ij}. These results will be used in the hypothesis tests presented in this paper. Both models (1) and (2) are discussed by Venter (1985). He derives the point estimators given below, but uses a different technique.

The derivation of the point estimators and their distributional properties is based on a technique from Zehnwirth (1984). The first step is to estimate θ_{ij} from the regression model for Y_{ijt}. We have

$$\hat{\theta}_{ij} = \sum_t P_{ijt}Y_{ijt}/P_{ij}, \; P_{ij} = \sum_t P_{ijt},$$

$$\hat{\theta}_{ij} \mid \theta_{ij} \sim (\theta_{ij}, V/P_{ij}),$$

and

$$\hat{\theta}_{ij} \mid \beta_i \sim N(\beta_i, 1/g_{ij} = V/P_{ij} + U_i).$$

Next, estimate β_i from the regression model for $\hat{\theta}_{ij}$:

$$\hat{\beta}_i = \sum_j g_{ij}\hat{\theta}_{ij}/g_i, \quad g_i = \sum_j g_{ij},$$

$$\hat{\beta}_i \mid \beta_i \sim N(\beta_i, 1/g_i),$$

and

$$\hat{\beta}_i \mid \mu \sim N(\mu, 1/d_i = 1/g_i + A).$$

Finally, estimate μ from the regression model for $\hat{\beta}_i$:

$$\hat{\mu} = \sum_i d_i\hat{\beta}_i/d, \; d = \sum_i d_i,$$

and

$$\hat{\mu} \mid \mu \sim N(\mu, 1/d).$$

If the normality assumption is removed, all of the moments remain correct and the estimators are best linear unbiased but not maximum likelihood.

Credibility estimators are then formed (again following Zehnwirth, 1984) by weighting the regression estimates by the ratio of variances:

$$\tilde{\beta}_i = W_i\hat{\beta}_i + (1 - W_i)\hat{\mu}, \quad W_i = Ad_i;$$
$$\tilde{\theta}_{ij} = Z_{ij}\hat{\theta}_{ij} + (1 - Z_{ij})\tilde{\beta}_i, \quad Z_{ij} = U_i g_{ij}.$$

It is worth noting that this approach can be used any time the observations can be written as a linear model. One example of this is the model with trend given by Hachemeister (1975).

In order to obtain either the regression estimates or the credibility estimates, it is necessary to estimate the variance $V, U_1, \ldots, U_k$, and A. The customary empirical Bayes approach is to postulate an appropriate sum of squares and then adjust it to produce an unbiased estimator. There is no problem in the estimation of V, the standard regression estimate being available. It is

$$R = \sum_{ijt} P_{ijt}(Y_{ijt} - \hat{\theta}_{ij})^2/(N - M).$$

We also have that $(N - M)R/V$ has a chi-square distribution with $N - M$ degrees of freedom.

The "natural" sums of squares to use for the other estimators would be the variance-weighted sums of squares. For U_i, that would be

$$S_i = \sum_j Z_{ij}(\hat{\theta}_{ij} - \hat{\beta}_i)^2/(m_i - 1),$$

and for A,

$$T = \sum_i W_i(\hat{\beta}_i - \hat{\mu})^2/(k - 1).$$

These are pseudo-estimators (De Vylder, 1981) in that the estimators depend on the quantities being estimated. They must be obtained by iteration using the current values on the right-hand side to obtain the updated values. Sometimes there is no positive solution (zero is always a solution) but this is a problem with all proposed estimators. When evaluating these estimators it is customary to assume that the variances in the formula are fixed at their true values. In that case, S_i and T are unbiased with $(m_i - 1)S_i/U_i$ and $(k-1)T/A$ having chi-square distributions with $m_i - 1$ and $k - 1$ degrees of freedom. In addition, $R, S_1, \ldots, S_k$, and T are mutually independent. All of these statements are shown by Klugman (1985a).

A competing estimator is derived from the literature on variance components models. Here weights are always equal to exposures so that the estimators are not based on unknown parameters. The sums of squares are

$$C_i = \sum_j P_{ij}(\hat{\theta}_{ij} - \beta_i^*)^2/(m_i - 1)$$

and

$$D = \sum_i P_i(\beta_i^* - \mu^*)^2/(k-1),$$

where $\beta_i^* = \sum_j P_{ij}\hat{\theta}_{ij}/P_i$, $\mu^* = \sum_i P_i\beta_i^*/P$, $P_{ij} = \sum_t P_{ijt}$, $P_i = \sum_j P_{ij}$, and $P = \sum_i P_i$. The unbiased estimators are then

$$S_i^* = (m_i - 1)(C_i - R)/(P_i - \sum_j P_{ij}^2/P_i)$$

and

$$T^* = [(k-1)(D-R) - \sum_i S_i^*(1/P_i - 1/P)\sum_j P_{ij}^2]/(P - \sum_i P_i^2/P).$$

These estimates are unbiased and $R, C_1, \ldots, C_k$ are mutually independent. However, D is not independent of the other estimators. If standardized in an appropriate way, $C_1, \ldots, C_k$ and D have approximate chi-square distributions. This gives $S_1, \ldots, S_k$ approximate chi-square distributions also. The above facts are all developed by Klugman (1985a).

The next two sections are concerned with confidence intervals for θ_{ij} based on $\tilde{\theta}_{ij}$, and hypothesis tests on the variances.

3. VARIANCES OF THE CREDIBILITY ESTIMATORS

If the structural parameters are known, it is relatively easy to obtain the desired variances. We have

$$\begin{aligned}\tilde{\theta}_{uv} &= Z_{uv}\hat{\theta}_{uv} + (1 - Z_{uv})[W_u\hat{\beta}_u + (1 - W_u)\hat{\mu}] \\ &= Z_{uv}\hat{\theta}_{uv} + (1 - Z_{uv})[W_u\sum_j g_{uj}\hat{\theta}_{uj}/g_u + (1 - W_u)\sum_{ij} d_i g_{ij}\hat{\theta}_{ij}/(g_i d)].\end{aligned}$$

The coefficient of $\hat{\theta}_{ij}$ is

$$\begin{aligned}&\delta(i,u)\delta(j,v)Z_{uv}+\delta(i,u)(1-Z_{uv})W_u g_{uj}/g_u\\&\quad+(1-Z_{uv})(1-W_u)d_i g_{ij}/(g_i d)\\&\quad=\delta(i,u)\delta(j,v)Z_{uv}+\delta(i,u)f_j+h_{ij},\end{aligned}$$

where $\delta(x,y)=1$ if $x=y$, and is zero otherwise. We also have (unconditionally), using model (2) that

$$\mathrm{Cov}(\hat{\theta}_{ij},\hat{\theta}_{rs})=\delta(i,r)A+\delta(i,r)\delta(j,s)(U_i+V/P_{ij}).$$

Then,

$$\begin{aligned}\mathrm{Var}(\tilde{\theta}_{uv})&=\sum_{ijrs}[\delta(i,u)\delta(j,v)Z_{uv}+\delta(i,u)f_j+h_{ij}]\\&\quad\times[\delta(r,u)\delta(s,v)Z_{uv}+\delta(r,u)f_s+h_{rs}]\\&\quad\times[\delta(i,r)A+\delta(i,r)\delta(j,s)(U_i+V/P_{ij})]\\&=\sum_{ijs}[\delta(i,u)\delta(j,v)Z_{uv}+\delta(i,u)f_j+h_{ij}]\\&\quad\times[\delta(i,u)\delta(s,v)Z_{uv}+\delta(i,u)f_s+h_{is}]\\&\quad\times[A+\delta(j,s)(U_i+V/P_{ij}]\\&=A\{Z_{uv}^2+\sum_{js}f_jf_s+\sum_{ijs}h_{ij}h_{is}+2\sum_j Z_{uv}f_j\\&\quad+2\sum_j Z_{uv}h_{uj}+2\sum_{js}f_jh_{us}\}+Z_{uv}^2(U_u+V/P_{uv})\\&\quad+\sum_j f_j^2(U_u+V/P_{uj})\\&\quad+\sum_{ij}h_{ij}^2(U_i+V/P_{ij})+2Z_{uv}f_v(U_u+V/P_{uv})\\&\quad+2Z_{uv}h_{uv}(U_u+V/P_{uv})+2\sum_j f_jh_{uj}(U_u+V/P_{uj})\\&=Z_{uv}^2(A+1/g_{uv})+(1-Z_{uv})W_u(W_u+2Z_{uv}-Z_{uv}W_u)/d_u\\&\quad+(1-Z_{uv})(1-W_u)(1+Z_{uv}+W_u-Z_{uv}W_u)/d.\end{aligned}$$

To use the above expression, estimates of the parameters are needed, but that will always be the case.

We now attempt to derive an expression for $\mathrm{Var}(\tilde{\theta}_{uv})$ that recognizes that W_u and Z_{uv} are also estimated from the data. We will continue to assume that the weights used in obtaining $\hat{\beta}_u$ and $\hat{\mu}$ are fixed quantities. The development is based on Morris (1983a,b) and is an extension of Klugman (1985b) where the problem is solved for the basic credibility model.

The first part of the analysis involves obtaining the posterior distribution of the θ_{ij}. At this stage, we continue to condition on the variances, but assign to μ a uniform prior on the real line. In the following, any vector contains all of the values of that quantity. For example, $\boldsymbol{\theta}' = (\theta_{11}, \ldots, \theta_{km_k})$. Also, we leave the variances $(V, U_1, \ldots, U_k, A)$ out of the conditonal statements. Every random variable is being conditioned on these parameters. Begin with

$$\mu \mid \hat{\boldsymbol{\theta}} : f(\mu \mid \hat{\boldsymbol{\theta}}) \propto f(\hat{\boldsymbol{\theta}} \mid \mu) f(\mu).$$

The prior is $f(\mu) \propto 1$ and $\hat{\boldsymbol{\theta}} \sim N(\mu\mathbf{1}, \mathbf{H})$, where $\mathbf{1}$ is a column vector of 1's and $\mathbf{H}$ is the covariance matrix. $\mathbf{H}$ is block diagonal where the ith block has an A in every cell and $A + 1/g_{i_j}$ on the diagonal. Then

$$\begin{aligned} f(\mu \mid \hat{\boldsymbol{\theta}}) &\propto \exp[-(\hat{\boldsymbol{\theta}} - \mu\mathbf{1})'\mathbf{H}^{-1}(\hat{\boldsymbol{\theta}} - \mu\mathbf{1})/2] \\ &\propto \exp[-(\mu - \mathbf{1}'\mathbf{H}^{-1}\hat{\boldsymbol{\theta}}/\mathbf{1}'\mathbf{H}^{-1}\mathbf{1})^2\mathbf{1}'\mathbf{H}^{-1}\mathbf{1}/2] \end{aligned}$$

Therefore,

$$\begin{aligned} \mu \mid \hat{\boldsymbol{\theta}} &\sim N(\mathbf{1}'\mathbf{H}^{-1}\hat{\boldsymbol{\theta}}/\mathbf{1}'\mathbf{H}^{-1}\mathbf{1}, 1/\mathbf{1}'\mathbf{H}^{-1}\mathbf{1}) \\ &\sim N(\hat{\mu}, 1/d). \end{aligned}$$

We next evaluate

$$\begin{aligned} f(\boldsymbol{\theta} \mid \hat{\boldsymbol{\theta}}, \boldsymbol{\beta}, \mu) &\propto f(\hat{\boldsymbol{\theta}} \mid \boldsymbol{\theta}, \boldsymbol{\beta}, \mu) f(\boldsymbol{\theta} \mid \boldsymbol{\beta}, \mu) \\ &\propto \exp[-\sum_{ij}(\hat{\theta}_{ij} - \theta_{ij})^2 P_{ij}/2V - \sum_{ij}(\theta_{ij} - \beta_i)^2/2U_i] \\ &\propto \exp[-\sum_{ij}\theta_{ij}^2(P_{ij}/V + 1/U_i)/2 + \sum_{ij}\theta_{ij}(\hat{\theta}_{ij}P_{ij}/V + \beta_i/U_i)] \\ &\propto \exp\{-\sum_{ij}[\theta_{ij} - Z_{ij}\hat{\theta}_{ij} - (1 - Z_{ij})\beta_i]^2/[2(1 - Z_{ij})U_i]\}. \end{aligned}$$

Therefore,

$$\theta_{ij} \mid \hat{\theta}_{ij}, \boldsymbol{\beta}, \mu \sim N[(Z_{ij}\hat{\theta}_{ij} - (1 - Z_{ij})\beta_i, (1 - Z_{ij})U_i]$$

and they are conditionally independent. Next,

$$\begin{aligned} f(\boldsymbol{\beta} \mid \hat{\boldsymbol{\theta}}, \mu) &\propto f(\hat{\boldsymbol{\theta}} \mid \boldsymbol{\beta}, \mu) f(\boldsymbol{\beta} \mid \mu) \\ &\propto \exp[-\sum_{ij} (\hat{\theta}_{ij} - \beta_i)^2 g_{ij}/2 - \sum_i (\beta_i - \mu)^2/(2A)] \\ &\propto \exp[-\sum_i \beta_i^2 (1/A + g_i)/2 + \sum_i \beta_i (\mu/A + \sum_j g_{ij}\hat{\theta}_{ij})] \\ &\propto \exp\{-\sum_i [\beta_i - W_i\hat{\beta}_i - (1 - W_i)\mu]^2/[2(1 - W_i)A]\}. \end{aligned}$$

Therefore, the conditional moments are

$$\beta_i \mid \hat{\boldsymbol{\theta}}, \mu \sim N[W_i\hat{\beta}_i + (1 - W_i)\mu, (1 - W_i)A],$$

$$\begin{aligned} E(\theta_{ij} \mid \hat{\boldsymbol{\theta}}, \mu) &= E[E(\theta_{ij} \mid \hat{\boldsymbol{\theta}}, \boldsymbol{\beta}, \mu)] \\ &= E[Z_{ij}\hat{\theta}_{ij} + (1 - Z_{ij})\beta_i \mid \hat{\boldsymbol{\theta}}, \mu] \\ &= Z_{ij}\hat{\theta}_{ij} + (1 - Z_{ij})[W_i\hat{\beta}_i + (1 - W_i)\mu], \end{aligned}$$

and

$$\begin{aligned} \mathrm{Var}(\theta_{ij} \mid \hat{\boldsymbol{\theta}}, \mu) &= E[\mathrm{Var}(\theta_{ij} \mid \hat{\boldsymbol{\theta}}, \boldsymbol{\beta}, \mu)] + \mathrm{Var}[E(\theta_{ij} \mid \hat{\boldsymbol{\theta}}, \boldsymbol{\beta}, \mu)] \\ &= E[(1 - Z_{ij})U_i \mid \hat{\boldsymbol{\theta}}, \mu] + \mathrm{Var}[Z_{ij}\hat{\theta}_{ij} + (1 - Z_{ij})\beta_i \mid \hat{\boldsymbol{\theta}}, \mu] \\ &= (1 - Z_{ij})U_i + (1 - Z_{ij})^2(1 - W_i)A. \end{aligned}$$

The expectation is the solution to the Bayesian version of this problem, and the variance is the posterior variance for constructing Bayesian credibility intervals. But we do not know μ and instead make use of the distribution of $\mu|\hat{\boldsymbol{\theta}}$ derived earlier. The moments are

$$\begin{aligned} E(\theta_{ij} \mid \hat{\boldsymbol{\theta}}) &= E[E(\theta_{ij} \mid \hat{\boldsymbol{\theta}}, \mu)] \\ &= E\{Z_{ij}\hat{\theta}_{ij} + (1 - Z_{ij})[W_i\hat{\beta}_i + (1 - W_i)\mu] \mid \hat{\boldsymbol{\theta}}\} \\ &= Z_{ij}\hat{\theta}_{ij} + (1 - Z_{ij})[W_i\hat{\beta}_i + (1 - W_i)\hat{\mu}], \end{aligned}$$

and

$$\begin{aligned} \mathrm{Var}(\theta_{ij} \mid \hat{\boldsymbol{\theta}}) &= E[\mathrm{Var}(\theta_{ij} \mid \hat{\boldsymbol{\theta}}, \mu)] + \mathrm{Var}[E(\theta_{ij} \mid \hat{\boldsymbol{\theta}}, \mu)] \\ &= E[(1 - Z_{ij})U_i + (1 - Z_{ij})^2(1 - W_i)A \mid \hat{\boldsymbol{\theta}}] \\ &\quad + \mathrm{Var}\{Z_{ij}\hat{\theta}_{ij} + (1 - Z_{ij})[W_i\hat{\beta}_i + (1 - W_i)\mu] \mid \hat{\boldsymbol{\theta}}\} \\ &= (1 - Z_{ij})U_i + (1 - Z_{ij})^2(1 - W_i)A \\ &\quad + (1 - Z_{ij})^2(1 - W_i)^2/d. \end{aligned}$$

The expectation is the usual credibility estimator given in Section 2. The variance takes on a simpler form than that given earlier in this section.

At this time we note that the Bühlmann-Straub (1972) model is indeed a special case of the hierarchical model with $A = 0$ and $U_i = U$ for all i. Then, the M classes are indistinguishable despite being doubly indexed. The variance components model becomes $Y_{ijt} = \mu + b_{ij} + e_{ijt}$ with each b_{ij} having variance U. The posterior conditional mean and variance are, respectively,

$$Z_{ij}\hat{\theta}_{ij} + (1 - Z_{ij})\hat{\mu}$$

and

$$(1 - Z_{ij})U + (1 - Z_{ij})^2/d.$$

They agree with the results given by Klugman (1985b).

As of now, little has been accomplished. In order to obtain an empirical Bayes solution, two more steps need to be taken. First, estimators of the unknown variances must be selected and their distributions (which will depend on the unknown variances) determined. Then, vague priors are placed on the unknown variances and their posterior distributions, determined using the model just obtained. Finally, these posterior distributions are used to remove the variances from the conditional expectation and variance given above. This approach is much closer to the pure Bayesian analysis. By employing estimators of the variances, it is reasonable to use vague priors, the data being given most of the weight. This method is different from that used by Zehnwirth (1982), where the unbiased estimates of the variances are modified by an additional credibility procedure.

The analysis is much easier when the sample sizes in each class are identical. That is, $P_{ij} = p_i$ for all i, j. In this case $1/g_{ij} = V/p_i + U_i$ depends only on i and so, Z_{ij} can be written as $U_i/(V/p_i + U_i)$. We then note that sums of squares S_i and C_i are related by

$$C_i = S_i(V + p_iU_i)/U_i$$

and $(m_i - 1)C_i/(V + p_iU_i)$ has a chi-square distribution with $m_1 - 1$ degrees of freedom. Unfortunately, T and D are not related in a similar manner.

Let $Q_i = V + p_iU_i$. Then,

$$R \mid V, Q, A \sim \Gamma[(N - M)/2, (N - M)/2V],$$
$$C_i \mid V, Q, A \sim \Gamma[(m_i - 1)/2, (m_i - 1)/2Q_i],$$

and

$$T \mid V, Q, A \sim \Gamma[(k - 1)/2, (k - 1)/2A],$$

where $\Gamma(a,b)$ indicates a gamma distribution with p.d.f. $b^a e^{-bx} x^{a-1}/\Gamma(a)$. The random variables $R, C_1, \ldots, C_k, T$ are mutually independent. The next step is to assign independent vague priors to V, $Q_1, \ldots, Q_k$, and A. A reasonable choice for a vague prior on the positive real numbers is $f(x) \propto 1/x$, $x > 0$. One indication of the vague nature of this prior is that it is the only one of the form x^a that has infinite area on both sides of any point. See Klugman (1985b) for a discussion of the effects of varying the exponent in the prior p.d.f. in the basic model. Morris (1983a) selected the exponents to produce an unbiased estimator of the credibility factor. That is not attempted here. One drawback of these priors is that they do not reflect the fact that $Q_i > V$. By placing prior probability on all positive Q_i, V pairs, probability is being assigned to negative values of U_i. This corresponds to the fact that estimator S_i^* can produce negative values. This problem is eliminated in the unequal sample size case at the expense of a tremendous increase in computation.

We then have,

$$\begin{aligned} &f(V,\mathbf{Q},A \mid R,\mathbf{C},T) \\ &\quad \propto f(R,\mathbf{C},T \mid V,\mathbf{Q},A) f(V,\mathbf{Q},A) \\ &\quad \propto \exp[-(N-M)R/2V - \sum_i (m_i - 1)C_i/2Q_i - (k-1)T/2A] \\ &\qquad \times V^{-(N-M+2)/2} A^{-(k+1)/2} \prod_i Q_i^{-(m+1)/2} \end{aligned}$$

or

$$\begin{aligned} V \mid R,\mathbf{C},T &\sim I\Gamma[(N-M)/2, (N-M)R/2], \\ Q_i \mid R,\mathbf{C},T &\sim I\Gamma[(m_i-1)/2, (m_i-1)C_i/2], \end{aligned}$$

and

$$A \mid R,\mathbf{C},T \sim I\Gamma[(k-1)/2, (k-1)T/2],$$

where $I\Gamma(a,b)$ is the inverse gamma distribution with p.d.f. $\propto e^{-b/x} x^{-a-1}$ and the random variables are conditionally independent. We are now prepared to obtain the empirical Bayes point estimator.

$$\begin{aligned} \theta_{ij}^{EB} &= E(\theta_{ij} \mid \hat{\boldsymbol{\theta}}, R, \mathbf{C}, T) \\ &= E[E(\theta_{ij} \mid \hat{\boldsymbol{\theta}}, R, \mathbf{C}, T, V, \mathbf{Q}, A)] \\ &= E\{Z_{ij} \mid \hat{\boldsymbol{\theta}}_{ij} + (1 - Z_{ij})[W_i \hat{\beta}_i + (1 - W_i)\hat{\mu}] \mid R, \mathbf{C}, T\} \\ &= E(Z_{ij} \mid R,\mathbf{C},T)(\hat{\theta}_{ij} - \hat{\mu}) + E(W_i \mid R,\mathbf{C},T)(\hat{\beta}_i - \hat{\mu}) \\ &\quad + E(Z_{ij} W_i \mid R,\mathbf{C},T)(\hat{\mu} - \hat{\beta}_i) + \hat{\mu} \\ &= Z_{ij}^{EB}(\hat{\theta}_{ij} - \hat{\mu}) + W_i^{EB}(\hat{\beta}_i - \hat{\mu}) + (Z_{ij}W_i)^{EB}(\hat{\mu} - \hat{\beta}_i) + \hat{\mu}. \end{aligned}$$

Since $Z_{ij} = p_i U_i / Q_i = 1 - V/Q_i$,

$$\begin{aligned} Z_{ij}^{EB} &= 1 - E(V \mid R, \mathbf{C}, T) E(Q_i^{-1} \mid R, \mathbf{C}, T) \\ &= 1 - [(N - M)/(N - m - 2)] R/C_i. \end{aligned}$$

Also, $W_i = A d_i = m_i p_i A/(Q_i + m_i p_i A)$. Then,

$$1 - W_i^{EB} = E[Q_i/(Q_i + m_i p_i A) \mid R, \mathbf{C}, T].$$

To find this expectation we use the following general result that is derived in the Appendix.

Result 1. If V, Q_i, and A have the conditional distributions given above, then writing m in place of m_i, we have that

$$E(V^s \mid R, \mathbf{C}, T) = [(N - M)R]/(N - M - 2)(N - M - 4) \ldots (N - M - 2s)$$

and that

$$\begin{aligned} &E[A^s Q_i^t (Q_i + rA)^u \mid R, \mathbf{C}, T] \\ &\quad = [(k-1)T]^s [(m-1)C_i]^{t+u} \Gamma[(m - 1 - 2t - 2u)/2] \Gamma[(k - 1 - 2s)/2] \\ &\quad \times E[(1 + rcF)^u]/\{2^{s+t+u} \Gamma[(m-1)/2] \Gamma[(k-1)/2]\}, \end{aligned}$$

where F has an F distribution, that is $F \sim F(m - 1 - 2t - 2u,\ k - 1 - 2s)$ and

$$c = [(k-1)(m - 1 - 2t - 2u)T]/[(m-1)(k - 1 - 2s)C_i].$$

The expectation must be evaluated numerically.

Let $e(s, t, u, r) = E[A^s Q_i^t (Q_i + rA)^u \mid R, \mathbf{C}, T]$. Then

$$\begin{aligned} 1 - W_i^{EB} &= e(0, 1, -1, m_i p_i) \\ &= E[(1 + m_i p_i T F/C_i)^{-1}], \end{aligned}$$

where $F \sim F(m_i - 1, k - 1)$.

Next we have

$$\begin{aligned} (Z_{ij} W_i)^{EB} &= E(Z_{ij} W_i \mid R, \mathbf{C}, T) \\ &= E\{(1 - V/Q_i)[1 - Q_i/(Q_i + m_i p_i A)] \mid R, \mathbf{C}, T\} \\ &= 1 - (1 - Z_{ij}^{EB}) - (1 - W_i^{EB}) + E[V/(Q_i + m_i p_i A) \mid R, \mathbf{C}, T] \end{aligned}$$

and

$$\begin{aligned}E[V/(Q_i m_i p_i A) \mid R, \mathbf{C}, T] &= [(N-M)R/(N-M-2)]e(0,0,-1,m_i p_i) \\ &= [(N-M)R/(N-M-2)C_i] \\ &\quad \times E\{[1 + m_i p_i (m_i+1)TF/((m_i-1)C_i)]^{-1}\},\end{aligned}$$

where $F \sim F(m_i+1, k-1)$.

We now do the same for the empirical Bayes variance. Let V_{ij}^{EB} be the estimate of the variance. That is

$$\begin{aligned}V_{ij}^{EB} &= \mathrm{Var}(\theta_{ij} \mid \hat{\boldsymbol{\theta}}, R, \mathbf{C}, T) \\ &= E[\mathrm{Var}(\theta_{ij} \mid \hat{\boldsymbol{\theta}}, R, \mathbf{C}, T, V, \mathbf{Q}, A)] + \mathrm{Var}[E(\theta_{ij} \mid \hat{\boldsymbol{\theta}}, R, \mathbf{C}, T, V, \mathbf{Q}, A)] \\ &= E[(1-Z_{ij})U_i + (1-Z_{ij})^2(1-W_i)A \\ &\quad + (1-Z_{ij})^2(1-W_i)^2/d \mid R, \mathbf{C}, T] \\ &\quad + \mathrm{Var}\{Z_{ij}\hat{\theta}_{ij} + (1-Z_{ij})[W_i\hat{\beta}_i + (1-W_i)\hat{\mu}] \mid R, \mathbf{C}, T\} \\ &= (*) + (**).\end{aligned}$$

Recall that $Z_{ij} = 1-V/Q_i$, $U_i = (Q_i - V)/p_i$, and $W_i = 1 - Q_i/(Q_i + m_i p_i A)$. Let $r = m_i p_i$, and $h(s) = E(V^s \mid R, \mathbf{C}, T)$, then,

$$\begin{aligned}(*) &= h(1)/p_i - h(2)e(0,-1,0,0)/p_i + h(2)e(1,-1,-1,r) \\ &\quad + h(2)e(0,0,-2,r)/d \\ &= \frac{(N-M)R}{(N-M-2)p_i} + \frac{[(N-M)R]^2}{(N-M-2)(N-M-4)} \\ &\quad \times \left\{ -\frac{1}{C_i p_i} + \frac{(k-1)(m_i+1)}{(k-3)(m_i-1)C_i^2} E\left\{\left[1 + \frac{r(k-1)(m_i+3)TF_1}{(m_i-1)(k-3)C_i}\right]^{-1}\right\}\right. \\ &\quad \left. + \frac{(m_i+1)}{(m_i-1)C_i^2} E\left\{\left[1 + \frac{r(m_i+3)TF_2}{(m_i-1)C_i}\right]^{-1}\right\}\right\},\end{aligned}$$

where $F_i \sim F(m_i+3, k-3)$ and $F_2 \sim F(m_i+3, k-1)$. We treat d as a constant even though it clearly contains random elements. This is the first of many approximations that have to be made in order to keep the computations remotely feasable. We complete the evaluation of $(*)$ by replacing d by its point estimator.

Next, we have that

$$\begin{aligned}(**) &= E\{[Z_{ij}\hat{\theta}_{ij} + (1 - Z_{ij})W_i\hat{\beta}_i + (1 - Z_{ij})(1 - W_i)\hat{\mu}]^2 \mid R, \mathbf{C}, T\} \\ &\quad - (\theta_{ij}^{EB})^2 \\ &= (\hat{\theta}_{ij})^2[1 - 2h(1)e(0,1,0,0) + h(2)e(0,2,0,0)] \\ &\quad + (\hat{\beta}_i)^2 r^2 h(2)e(2,-2,-2,r) + \mu^2 h(2)e(0,0,-2,r) \\ &\quad + 2\hat{\theta}_{ij}\hat{\beta}_i r[h(1)e(1,-1,-1,r) - h(2)e(1,-2,-1,r)] \\ &\quad + 2\hat{\theta}_{ij}\hat{\mu}[h(1)e(0,0,-1,r) - h(2)e(0,1,-1,r)] \\ &\quad + 2\hat{\beta}_i\hat{\mu}rh(2)e(1,-1,-2,r) - (\theta_{ij}^{EB})^2.\end{aligned}$$

These formulas require a considerable amount of calculation. To obtain a simpler version, consider the case where all the sample sizes are equal, $P_{ij} = p$ for all i, j, all the groups have the same number of classes, $m_i = m$ for all i, and the within group variances are equal, $U_i = U$ for all i. We then have a pooled estimator

$$C = p\sum_{ij}(\hat{\theta}_{ij} - \hat{\beta}_i)^2/(M - k),$$

with $(M-k)C/(V+pU)$ having a chi-square distribution with $M-k$ degrees of freedom. We also have

$$D = mp\sum_{i}(\hat{\beta}_i - \hat{\mu})^2/(k - 1),$$

with $(k-1)D/(V + pU + mpA)$ having a chi-square distribution with $k-1$ degrees of freedom. In this setting, $\hat{\beta}_i = \beta_i^*$ and $\hat{\mu} = \mu^*$. In addition, R, C, and D are independent. Now let $Q = V + pU$ and $B = V + pU + mpA$. Then,

$$R \mid V, Q, B \sim \Gamma[(N - M)/2, (N - M)/2V],$$
$$C \mid V, Q, B \sim \Gamma[(M - k)/2, (M - k)/2Q],$$

and

$$D \mid V, Q, B \sim \Gamma[(k - 1)/2, (k - 1)/2B].$$

Using the same vague priors as before,

$$V \mid R, C, D \sim I\Gamma[(N - M)/2, (N - M)R/2],$$
$$Q \mid R, C, D \sim I\Gamma[(M - k)/2, (M - k)C/2],$$

and

$$B \mid R, C, D \sim I\Gamma[(k-1)/2, (k-1)D/2].$$

It is now easy to compute θ_{ij}^{EB}. We will use

$$\begin{aligned} Z_{ij}^{EB} &= E(pU/Q \mid R, C, D) \\ &= E(1 - V/Q \mid R, C, D) \\ &= 1 - (N-M)R/(N-M-2)C, \end{aligned}$$

$$\begin{aligned} W_i^{EB} &= E(pmA/B \mid R, C, D) \\ &= E(1 - Q/B \mid R, C, D) \\ &= 1 - (M-k)C/(M-k-2)D, \end{aligned}$$

and

$$\begin{aligned} (Z_{ij}W_i)^{EB} &= E[(1 - V/Q)(1 - Q/B) \mid R, C, D] \\ &= 1 - (1 - Z_{ij}^{EB}) - (1 - W_i^{EB}) + (N-M)R/(N-M-2)D \\ &= Z_{ij}^{EB} + W_i^{EB} + (N-M)R/(N-M-2)D - 1. \end{aligned}$$

For computing v_{ij}^{EB}, we next have, noting that $d = kmp/B$, that

$$\begin{aligned} (*) &= E(VU/Q + V^2A/QB + V^2/B^2d \mid R, C, D) \\ &= E[\frac{V}{p} - \frac{V^2}{pQ} + \frac{V^2}{mpQ} - \frac{V^2}{mpB} + \frac{V^2}{kmpB} \mid R, C, D] \\ &= \frac{(N-M)R}{(N-M-2)p} - \frac{(N-M)^2R^2}{(N-M-2)(N-M-4)}\left[\frac{m-1}{mpC} + \frac{k-1}{kmpD}\right] \end{aligned}$$

and that

$$\begin{aligned} (**) &= (\hat{\theta}_{ij})^2 E(1 - 2V/Q + V^2/Q^2 \mid R, \mathrm{C}, D) \\ &\quad + (\hat{\beta}_i)^2 E(V^2/Q^2 - 2V^2/QB + V^2/B^2 \mid R, \mathrm{C}, D) \\ &\quad + \hat{\mu}^2 E(V^2/B^2 \mid R, \mathrm{C}, D) \\ &\quad + 2\hat{\theta}_{ij}\hat{\beta}_i E(V/Q - V^2/Q^2 - V/B + V^2/QB \mid R, \mathrm{C}, D) \\ &\quad + 2\hat{\theta}_{ij}\hat{\mu} E(V/B - V^2/QB \mid R, \mathrm{C}, D) \\ &\quad + 2\hat{\beta}_i\hat{\mu} E(V^2/QB - V^2/B^2 \mid R, \mathrm{C}, D) - (\theta_{ij}^{EB})^2 \\ &= \frac{(N-M)^2R^2}{(N-M-2)(N-M-4)}\left[\frac{(M-k+2)(\hat{\theta}_{ij} - \hat{\beta}_i)^2}{(M-k)C^2}\right. \\ &\quad \left. + \frac{(k-3)(\hat{\beta}_i - \hat{\mu})^2}{(k-1)D^2} + \frac{2(\hat{\theta}_{ij} - \hat{\beta}_i)(\hat{\beta}_i - \hat{\mu})}{CD}\right] - (\hat{\theta}_{ij} - \hat{\theta}_{ij}^{EB})^2. \end{aligned}$$

Since no approximate integrations are needed, this is clearly more convenient. It can also be used as an approximation to the case discussed at the beginning of this section. To do so, substitute p_i for p, m_i for m, $C_i + m_i p_i T$ for D, and C_i for C. In addition, whenever the quantity $M - k$ appears, it should be replaced by $m_i - 1$.

Finally, in the case where the P_{ij} need not be equal at all, replacements are still possible. In the formulas at the beginning of this section, use P_{ij} for p_i, and $R + p_i S_i$ for C_i. To avoid approximate integration, use the formulas at the end of this section with P_{ij} for p, m_i for m, $R + P_{ij} S_i$ for C, $(R + P_{ij} S_i)(1 + T \sum_u \frac{P_{iu}}{R + P_{iu} S_i})$ for D, and $m_i - 1$ for $M - k$. It remains to be seen whether or not these approximations are satisfactory for use. Klugman (1985b) gave an example that indicated that when P_{ij} is small and θ_{ij} is far from μ the approximation is likely to be of little value. It performed very well in the remaining cases.

4. HYPOTHESIS TESTS

When considering a hierarchical model it would be useful to know if such a model is appropriate. In this section we present four hypothesis tests that can help in this decision process. The questions to be investigated are (1) Does the grouping reflect true differences? (2) Can data be pooled when estimating the within group variances? (3) Is the basic model of Bühlmann and Straub (1972) sufficient?

The four null hypotheses are

$$H1 : A = 0$$
$$H2 : U_1 = \ldots = U_k = U$$
$$H2' : U_1 = \ldots = U_k = 0$$
$$H3 : U_1 = U_2$$
$$H4 : \beta_1 = \beta_2$$

The last two are presented for the first two groups. The results clearly apply to any pair of groups. The following conclusions can be drawn when various combinations of these hypotheses are accepted.

1. Accept $H1$ and $H2$: The basic model can be used over the M classes. The grouping can be ignored.
2. Accept $H1$, reject $H2$: Use k separate basic models, one for each group.
3. Reject $H1$, accept $H2$: Use a pooled estimate of U but each group has its own estimated mean.

4. Reject $H1$, accept $H2$, accept $H2'$: Ignore the classes, only the groups produce significant differences.
5. Reject $H1$, reject $H2$, accept $H3$, accept $H4$: Groups 1 and 2 can be combined.
6. Reject $H1$, reject $H2$, accept $H3$, reject $H4$: Groups 1 and 2 can be pooled when estimating the common value of U.
7. Reject $H1$, reject $H2$, reject $H3$, accept $H4$: Groups 1 and 2 can be pooled when estimating the common value of β.

To test $H1$ note that $A = 0$ is equivalent to $\beta_1 = \ldots = \beta_k$. This hypothesis can be testing by using an analysis of variance. The model to use is

$$\hat{\theta}_{ij} \mid \beta_i \sim N(\beta_i, 1/g_{ij})$$

and the test statistic is

$$\frac{\sum_i g_i(\hat{\beta}_i - \hat{\beta})^2/(k-1)}{\sum_{ij} g_{ij}(\hat{\theta}_{ij} - \hat{\beta}_i)^2/(M-k)},$$

where g_{ij} is estimated by $(S_i + R/P_{ij})^{-1}$ and $\hat{\beta} = \sum_i g_i\hat{\beta}_i / \sum_i g_i$. Under $H1$, this ratio has an F distribution with $k-1$ and $M-k$ degrees of freedom. Reject $H1$ if the observed value is too large.

To test $H2$, we use Bartlett's test (Snedecor and Cochran, 1967). We have as the test statistic

$$\frac{(M-k)\log\{\sum_i(m_i-1)S_i/(M-k)\} - \sum_i(m_i-1)\log S_i}{1+[\sum_i 1/(m_i-1) - 1/(M-k)]/3(k-1)}.$$

Under the null hypothesis, the test statistic has an approximate chi-square distribution with $k-1$ degrees of freeedom. Reject $H2$ if the value is large. For the chi-square approximation to be valid, it is recommended that each m_i be at least 5. It should also be noted that this test is very sensitive to non-normality. With heavy-tailed data there will be more rejection than expected. In our case, that will encourage use of the hierarchical model in some cases where it is not needed.

If $H2$ has been accepted, then we might want to also test $H2'$. This is equivalent to testing $\theta_{11} = \ldots = \theta_{1m_1}, \ldots, \theta_{k1} = \ldots = \theta_{km_k}$. The appropriate F-ratio is

$$\frac{\sum_{ij} P_{ij}(\hat{\theta}_{ij} - \hat{\beta}_i)^2/(M-k)}{\sum_{ijt} P_{ijt}(Y_{ijt} - \hat{\theta}_{ij})^2/(N-M)} = \frac{\sum_i(m_i-1)C_i/(M-k)}{R}$$

and under $H2'$ it has an F distribution with $M-k$ and $N-M$ degrees of freedom.

The remaining two hypotheses can be considered as special cases of those tested above. For $H3$ use S_1/S_2. It has an F distribution with m_1-1 and m_2-1 degrees of freedom. Reject $H3$ if the ratio is large or small. For $H4$, use the two sample version of the test given in $H1$. The test statistic is the square root of

$$\frac{g_1 g_2(\beta_1-\beta_2)^2/(g_1+g_2)}{\sum_{ij} g_{ij}(\hat{\theta}_{ij}-\hat{\beta}_i)^2/(m_1+m_2-2)}.$$

Under $H4$, it has a t distribution with m_1+m_2-2 degrees of freedom. Reject for both large and small values. Of course, repeated testing of $H3$ and $H4$ on various pairs of groups will destory the overall significance level. Some protection will be needed if the level is to be maintained.

APPENDIX

Proof of Result 1. The first expectation follows directly from the inverse gamma distribution. The second can be written, dropping the subscript i on m and C, as

$$d\int_0^\infty\int_0^\infty a^s q^t(q+ra)^u a^{-(k+1)/2} q^{-(m+1)/2} \times \exp[-(k-1)T/2a-(m-1)C/2q]\,da\,dq,$$

where

$$d=\frac{[(m-1)C]^{(m-1)/2}[(k-1)T]^{(k-1)/2}}{\Gamma[(m-1)/2]\Gamma[(k-1)/2]2^{(m+k-2)/2}}.$$

Let $f=za/q$ where z is an arbitrary constant and q is the variable being transformed. The integral becomes

$$\int_0^\infty\int_0^\infty z^{(2t-m+1)/2} f^{(m-2t-3)/2} a^{(2s+2t+2u-m-k)/2} \times (z/f+r)^u \exp\{-[(m-1)Cf/2z+(k-1)T/2]/a\}\,df\,da.$$

Let $a = m - 2t - 2u - 1$ and $b = k - 2s - 1$. The integral is then

$$\int_0^\infty \int_0^\infty z^{(-a-2u)/2} f^{(a+2d-2)/2} a^{-(a+b+2)/2} (z/f + r)^u$$
$$\times \exp\{-[(m-1)Cf/2z + (k-1)T/2]/a\} da df$$
$$= \int_0^\infty \int_0^\infty z^{(-a-2u)/2} f^{(a-2)/2} a^{-(a+b+2)/2} (z + rf)^u$$
$$\times \exp\{-[(m-1)Cf/2z + (k-1)T/2]/a\} da df$$
$$= \int_0^\infty z^{(-a-2u)/2} f^{(a-2)/2} [(a+b)/2]$$
$$\times [(m-1)Cf/2z + (k-1)T/2]^{-(a+b)/2} (z + rf)^u df$$
$$= \Gamma[(a+b)/2] z^{-a/2} [2b/a(k-1)T]^{(a+b)/2}$$
$$\times \int_0^\infty f^{(a-2)/2} \left[\frac{(m-1)Cbf}{za(k-1)T} + \frac{b}{a}\right]^{-(a+b)/2} \left(1 + \frac{rf}{z}\right)^u df.$$

Letting $z = (m-1)Cb/[a(k-1)T]$, the integrand becomes (inserting appropriate constants to obtain the p.d.f. $f(f)$ of a random variable with an F distribution with a and b degrees of freedom)

$$[(m-1)C]^{-a/2} [(k-1)T]^{-b/2} 2^{(a+b)/2} \int_0^\infty f(f)(1 + rf/z)^u \Gamma(a/2)\Gamma(b/2) df.$$

Inserting the coefficient d in front of the integral produces the desired result.

REFERENCES

Bühlmann, H. (1967), "Experience rating and credibility". *ASTIN Bulletin* **4**, 199–207.

Bühlmann, H., and E. Straub (1972), "Credibility for loss ratios". *ARCH* **1972.2.**

De Vylder, F. (1981), "Practical credibility theory with emphasis on optimal parameter estimation". *ASTIN Bulletin* **12**, 115–131.

Gerber, H. (1982), "An unbayesed approach to credibility". *Insurance: Mathematics and Economics* **1**, 271–276.

Graybill, F. (1961), *An Introduction to Linear Statistical Models*, Volume 1. New York: McGraw-Hill.

Hachemeister, C. (1975), "Credibility for regression models with applications to trend". In *Credibility: Theory and Applications*, ed. P. M. Kahn, pp. 129–163. New York: Academic Press.

Insurance Services Office (1980), "Report of the Credibility Subcommittee: Development and testing of empirical Bayesian credibility procedures for classification ratemaking". New York: ISO.

Klugman, S. (1985a), "Distributional aspects and evaluation of some variance estimators in credibility models". *ARCH* **1985.1,** 73–97.

Klugman, S. (1985b), "Parametric empirical Bayes inference in credibility models". Unpublished manuscript.

Lindley, D., and A. Smith (1972), "Bayes estimates for the linear model". *Journal of the Royal Statistical Society,* Series B **34,** 1–41.

Morris, C. (1983a), "Parametric empirical Bayes confidence intervals". In *Proceedings of the Conference on Scientific Inference, Data Analysis, and Robustness,* ed. G. E. P. Box, T. Leonard, and C.-F. Wu, pp. 25–50. New York: Academic Press.

Morris, C. (1983b), "Parametric empirical Bayes inference: theory and applications". *Journal of the American Statistical Association* **78,** 47–55.

Snedecor, G. W., and W. G. Cochran (1967), *Statistical Methods.* Ames, Iowa: Iowa State University Press.

Venter, G. (1985), "Structured credibility in applications—hierarchical, multidimensional, and multivariate models", paper presented at the 19th Actuarial Research Conference, Berkeley, CA, to appear in *ARCH.*

Zehnwirth, B. (1982), "Conditional linear Bayes rules for hierarchical models". *Scandinavian Actuarial Journal,* 143–152.

Zehnwirth, B. (1984), "Credibility: estimation of structural parameters". In *Premium Calculation in Insurance,* ed. F. De Vylder, M. Goovaerts, and J. Haezendonck, pp. 347–359. Dordrecht, Holland: Reidel.

Harry H. Panjer [1]

MODELS OF CLAIM FREQUENCY

1. INTRODUCTION

One of the primary goals of actuarial risk theory is the evaluation of the risk associated with a portfolio of insurance contracts over the life of the contracts. Many insurance contracts (in both life and non-life areas) are short-term, typically one year. Typical automobile insurance, homeowner's insurance and group life and health insurance policies are usually of a one year duration.

One of the primary objectives of risk theory is to model the distribution of total claim costs for portfolios of policies so that business decisions can be made regarding various aspects of the insurance contracts. The total claim cost over a fixed time period is often modelled by considering the frequency of claims and the sizes of the individual claims separately. Hogg and Klugman (1984) considered a number of different possible claim size distributions and discussed estimation procedures for them.

In this paper, we are concerned with modelling the number of claims that arise from a portfolio of insured risks. Furthermore we are motivated by the ultimate exercise of obtaining numerical values for the distribution of total claim costs.

2. MATHEMATICAL BACKGROUND

Let $X_1, X_2, X_3, \ldots$ be independent and identically distributed random variables with common distribution function $F_X(x)$. Let N denote the number of claims occurring in a fixed time period. Assume that the distribution of each X_i, $i = 1, \ldots, N$, is independent of N for fixed N. Then the total claim cost for the fixed time period can be written as

$$S = X_1 + X_2 + \cdots + X_N$$

[1] Department of Statistics and Actuarial Science, University of Waterloo, Waterloo, Ontario N2L 3G1

I. B. MacNeill and G. J. Umphrey (eds.), Actuarial Science, 115–125.

with distribution function

$$F_S(x) = \sum_{n=0}^{\infty} p_n F_X^{*n}(x) \tag{1}$$

where $F_X^{*n}(\cdot)$ indicates the n-fold convolution of $F_X(\cdot)$.

Panjer (1981) showed that if the claim frequency distribution satisfies

$$p_n = \left(a + \frac{b}{n}\right) p_{n-1}, \qquad n = 1, 2, 3, \ldots \tag{2}$$

and that if the claim size distribution is discrete with support only on the positive integers, then the distribution of total claims can be calculated recursively as

$$f_S(x) = \sum_{y=1}^{x} \left\{a + b\frac{y}{x}\right\} f_X(y) f_S(x-y), \qquad x = 1, 2 \ldots. \tag{3}$$

It was shown that the number of computations involved in computing $f_S(x), x = 0, 1, \ldots, k$ is of order k^2 using (3) and of order k^3 using (1). For large k, the recursive relation (3) offers a computing tool that is much more efficient than (1). Panjer (1981) also gave the continuous analogue of (3), namely

$$f_S(x) = p_1 f_X(x) + \int_0^x \left\{a + b\frac{y}{x}\right\} f_X(y) f_S(x-y) dy, \quad x > 0.$$

Panjer (1981) showed that recursion (2) was satisfied by the binomial, Poisson and negative binomial and geometric distributions. Sundt and Jewell (1981) showed that these are the only distributions satisfying (2). Furthermore, they showed that if (2) was satisfied only for $n = m+1, m+2, \ldots$, then (3) could be replaced by

$$\begin{aligned} f_S(x) = &\sum_{y=1}^{x} \left\{a + b\frac{y}{x}\right\} f_X(y) f_S(x-y) \\ &+ \sum_{n=1}^{m} \left\{p_n - \left(a + \frac{b}{n}\right) p_{n-1}\right\} f_X^{*n}(x), \qquad x > 0, \end{aligned} \tag{4}$$

which involves at most m-fold convolutions of $f_X(\cdot)$. An important case is $m = 1$, with $p_0 = 0$ yielding

$$f_S(x) = \sum_{y=1}^{x} \left\{a + b\frac{y}{x}\right\} f_X(y) f_S(x-y) + p_1 f_X(x). \tag{5}$$

It is easily shown that the logarithmic distribution satisfies (2) only from $n = 2, 3, \ldots$ with $p_0 = 0$. It is a special case with $0 < a = -b < 1$.

Let $L_X(z) = E[e^{zX}]$ denote the Laplace transform (LT) of the random variable X (or its distribution). Let $P_N(z) = E[z^N]$ denote the probability generating function (pgf) of the random variable N (or its distribution). Then the LT of the total claim cost S satisfies

$$L_S(z) = P_N[L_X(z)]. \tag{6}$$

Panjer and Willmot (1984) have given a useful result for situations in which X is a nonnegative random variable with an atom (spike or point mass) at $x = 0$. If the probability generating function of N involves some parameter θ so that $P_N(z)$ can be written as

$$P_N(z;\theta) = B[\theta(z-1)] \tag{7}$$

for some function $B(\cdot)$, then the LT of S given by (6) can be rewritten as

$$L_S(z) = P_N[L_{X_+}(z); (1 - f_0)\theta] \tag{8}$$

where $L_{X_+}(z)$ is the LT of the conditional random variable $X_+ = X \mid X > 0$. Many distributions including the ones mentioned previously satisfy equation (7).

3. CONTAGIOUS DISTRIBUTIONS

Let $P_1(z)$ and $P_2(z)$ denote the pgf's of two discrete distributions defined on the positive integers. Then the distribution with pgf $P(z) = P_1(P_2(z))$ is said to be a contagious (or generalized or stopped) distribution. Douglas (1980) and Johnson and Kotz (1969) discuss some contagious distributions and their applications. It is possible to generalize to three or more component distributions but it will turn out to be unnecessary here.

Contagious distributions have natural applications in the modelling of insurance claims frequencies. Suppose that in some fixed time period (such as one year), M claim-causing events or "accidents" occur. Further, suppose that N_j claims arise from the jth accident. Then the total number of claims can be written as $N = N_1 + N_2 + \cdots + N_M$. Under the assumptions that for a fixed number of accidents the number of claims arising from each of the accidents are independent of each other and independent of N, the pfg of the total number of claims has pgf $P_N(z) = P_1(P_2(z))$ where $P_1(z)$ is the pgf of M and $P_2(z)$ is the pgf of the N_j's. From equation (6) the LT of total claim costs is

$$L_S(z) = P_1(P_2(L_X(z))). \tag{9}$$

Equation (9) suggests that computing the distribution of S can be done in two stages. First the distribution with Laplace transform $L(z) = P_2(L_X(z))$ can be calculated; then the distribution of S with LT $L_S(z) = P_1(L(z))$ can be calculated. A well-known result in actuarial science can be used to illustrate these points. Suppose that the number of accidents has a Poisson distribution with pgf $P_1(z) = \exp\{\lambda(z-1)\}$ and that the numbers of claims per accident have a common logarithmic distribution with pgf

$$P_2(z) = \{\log(1-qz)\}/\log(1-q).$$

Then the total number of claims has pgf

$$\begin{aligned} P_N(z) &= P_1(P_2(z)) \\ &= \exp\left\{\lambda\left[\frac{\log(1-qz)}{\log(1-q)} - 1\right]\right\} \\ &= \{1-\beta(z-1)\}^{-r}, \end{aligned} \tag{10}$$

where $\beta = q/(1-q)$ and $r = -\lambda/\log(1-q)$. The final expression in (10) is the pgf of a negative binomial distribution. Consequently, the distribution of total claim costs can be calculated in two different ways. First, it can be done directly using (3) with the values of a and b from the negative binomial distribution. Alternatively (although impractical), it could also be calculated in two steps. First, one calculates the distribution with LT $L(z) = P_2(L_X(z))$, where $P_2(z)$ is the pgf of the logarithmic distribution using (5) with a and b from the logarithmic distribution. Then one calculates the distribution with LT $L_S(z) = P_1(L(z))$ using (3) with a and b from the Poisson distribution.

The negative binomial has frequently been used to model claim numbers. In a major study of California driver records, Weber (1970) showed that the negative binomial distribution provided an excellent fit to the number of claims experienced by automobile drivers over the period 1961–1963. The results for 1961, 1962 and part of 1963 (totaling 2.875 years) are given in Table 1. It is obvious that the distribution fits extremely well. Many similar data sets are available in the literature for which the negative binomial does not provide an adequate fit. In most cases the negative binomial gives a right-hand tail that is not heavy enough. This motivates us to find alternative models.

4. THE CLASS OF CLAIM FREQUENCY DISTRIBUTIONS

Consider the class of discrete distributions with pgf of the form

$$P(z) = P_1(P_2(z)),$$

Table 1. *Actual and Theoretical (Negative Binomial) Accident Distribution — 1964 California Driver Record Study*

Number of Accidents	Actual Distribution	Theoretical Distribution
0	122,593	122,638
1	21,350	21,257
2	3,425	3,457
3	530	550
4	89	86
5+	19	18
	148,006	148,006

where $P_1(\cdot)$ satisfies recurrence relation (2) for $n = 1, 2, 3, \ldots$ and $P_2(\cdot)$ satisfies recurrence relation (2) only for $n = 2, 3, 4, \ldots$. Since $P_1(\cdot)$ is of the form (7) for each member of the class satisfying (2) (Poisson, binomial, negative binomial and geometric), we restrict, without loss of generality, the pgf $P_2(\cdot)$ to correspond to distributions defined only on $n = 1, 2, \ldots$; i.e., with probability zero at $n = 0$. This restriction is convenient from a computational point of view as well as from the point of view of interpreting the model. As in the Poisson-logarithmic example contained in the previous section, it is useful to interpret $P_1(z)$ as the pgf of the distribution of the number of "accidents" with at least one claim in the time period and $P_2(z)$ as the number of claims per accident. Then $P(z) = P_1(P_2(z))$ is the pgf of the total number of claims for all accidents in the time period.

The members of the class of distributions with pgf $P_2(z)$ are the (zero) truncated versions of the Poisson, binomial, negative binomial and geometric distributions as well as the logarithmic distribution and the *extended truncated negative binomial* distribution with pgf

$$P(z) = \frac{[1-\beta(z-1)]^{-r} - (1+\beta)^{-r}}{1-(1+\beta)^{-r}}, \quad \beta > 0, \quad -1 < r < 0. \tag{11}$$

This pgf has the same form as for the truncated negative binomial except that the range of r is extended to negative binomial values (see formula (10)). From Sundt and Jewell (1981), one can show that these are the only members of the class of distributions defined on $x = 1, 2, 3, \ldots$ satisfying recurrence relation (2) for $n = 2, 3, \ldots$.

The class of distributions with pgf $P(z) = P_1(P_2(z))$ with $P_1(z)$ and $P_2(z)$ as described above provides a rich class of contagious distributions with two, three or four parameters which can be used to fit claim frequency data.

5. SOME MEMBERS OF THE CLASS

Since the Poisson, binomial, negative binomial and geometric distributions are all of the form (7), it is not necessary to specify whether the distribution with $P_2(z)$ is truncated at zero. Consequently, we will refer only to the untruncated distribution except in the case of the logarithmic and extended truncated negative binomial distributions for which truncation is necessary.

5.1 Poisson-Binomial Distribution

The pgf of the Poisson distribution is

$$P_1(z) = \exp\{\lambda(z-1)\}, \qquad \lambda > 0 \tag{12}$$

and that of the binomial distribution is

$$P_2(z) = [1 + q(z-1)]^n, \qquad 0 < q < 1. \tag{13}$$

The resulting distribution with pgf $P(z) = P_1(P_2(z))$ is a three-parameter distribution with mean, variance and coefficient of a skewness given by

$$\mu_1' = n\lambda q,$$

$$\mu_2 = \mu_1'[1 + (n-1)q], \tag{14}$$

and

$$\mu_3\{\mu_2\}^{-\frac{3}{2}} = \mu_2^{-\frac{3}{2}}\left\{3\mu_2 - 2\mu_1' + \frac{(n-2)}{(n-1)}\frac{(\mu_2-\mu_1')^2}{\mu_1'}\right\}.$$

When $n = 1$, the binomial distribution becomes the two point Bernoulli distribution and the Poisson-Binomial distribution becomes the Poisson. Hence, it is only meaningful to consider $n \geq 2$. When $n = 1$ the distribution is the Hermite distribution. Using the interpretation used previously, at most two claims can occur per accident for the Hermite distribution. For fixed mean and variance, larger skewness results as n increases. However, since $(n-2)/(n-1)$ approaches 1 as n increases, the skewness is bounded.

5.2 Neyman Type A Distribution

This distribution has a Poisson distribution for both $P_1(z)$ and $P_2(z)$ so that

$$P_1(z) = \exp\{\lambda_1(z-1)\}, \qquad \lambda_1 > 0$$

and

$$P_2(z) = \exp\{\lambda_2(z-1)\}, \qquad \lambda_2 > 0.$$

The mean, variance and coefficient of skewness of the resulting two-parameter Neyman Type A distribution are

$$\mu_1' = \lambda_1\lambda_2,$$

$$\mu_2 = \mu_1'(1+\lambda_2), \tag{15}$$

and

$$\mu_3\{\mu_2\}^{-\frac{3}{2}} = \mu_2^{-\frac{3}{2}}\left\{3\mu_2 - 2\mu_1' + \frac{(\mu_2-\mu_1')^2}{\mu_1'}\right\}.$$

On comparing (14) and (15), it can be seen that for fixed mean and variance, the Neyman Type A distribution is the limiting case of the Poisson-Binomial distribution as n gets large. This is a direct consequence of the Poisson being a limiting form of the binomial. When λ_1 is small, the distribution is approximately a "Poisson with zeros" distribution with pgf

$$P(z) = \lambda_1 + (1-\lambda_1)^{\lambda_2(z-1)}.$$

When λ_2 is small, the distribution is approximately Poisson with mean $\lambda_1\lambda_2$.

5.3 Polya-Aeppli Distribution

This is a two-parameter contagious distribution with

$$P_1(z) = \exp\{\lambda(z-1)\}, \qquad \lambda > 0,$$

the Poisson distribution, and

$$P_2(z) = \{1-\beta(z-1)\}^{-1}, \qquad \beta > 0,$$

the geometric distribution. Like the previous two examples, its variance exceeds its mean, since

$$\mu_1' = \lambda\beta$$

and

$$\mu_2 = \mu_1'(1+2\beta).$$

The coefficient of skewness for the Polya-Aeppli distribution is

$$\mu_2^{-\frac{3}{2}} \left\{ 3\mu_2 - 2\mu_1' + \frac{3}{2} \frac{(\mu_2 - \mu_1')^2}{\mu_1'} \right\}. \tag{16}$$

Comparing (16) with the coefficient of skewness for the Newman Type A distribution given in (15), one sees that for fixed means and variance, this distribution is more skewed than the Neyman Type A distribution.

5.4 Generalized Poisson-Pascal Distribution

This three-parameter distribution has

$$P_1(z) = \exp\{\lambda(z-1)\}, \qquad \lambda > 0$$

the Poisson distribution and $P_2(z)$ the extended truncated negative binomial distribution of form (11) with $r > -1$. This distribution has mean, variance and coefficient of skewness

$$\mu_1' + \lambda[1 - (1+\beta)^{-r}]^{-1} r\beta,$$

$$\mu_2 = \mu_1'[1 + (r+1)\beta]$$

and

$$\mu_2^{-\frac{3}{2}} \left\{ 3\mu_2 - 2\mu_1' + \frac{r+2}{r+1} \frac{(\mu_2 - \mu_1')^2}{\mu_1'} \right\},$$

respectively. The negative binomial distribution is the special (limiting) case of $r = 0$. The Polya-Aeppli distribution is the special case with $r = 1$ and the Neyman Type A distribution is the limiting case as $r \to \infty$. A very important special case is the Poisson–Inverse Gaussian distribution with $r = -\frac{1}{2}$. It generalizes these two-parameter distributions at the expense of an additional parameter. This is an important consideration when testing the fit of a distribution to data as will be seen in the next section.

The Poisson–Inverse Gaussian derives its name from the fact that it can be viewed as a mixed Poisson distribution with inverse Gaussian mixing distribution (Willmot, 1986).

6. NUMERICAL ILLUSTRATIONS

Table 2 gives two sets of observed data for frequency of automobile claims. They are taken from Bühlmann (1970) and Hossack *et al.* (1983).

For the first data set a negative binomial distribution was fitted using maximum likelihood but the fit was judged to be inadequate based on a

Table 2. *Automobile Accident Frequency Data*

Number of Accidents	No. of Drivers Bühlmann	Hossack *et al.*
0	103,704	565,664
1	14,075	68,714
2	1,766	5,177
3	255	365
4	45	24
5	6	6
6	2	0
	119,853	639,950

chi-square goodness-of-fit statistic of 12.37 on 3 degrees of freedom for a significance level of .6%. When the Poisson–Inverse Gaussian distribution was fitted to the same data the chi-square statistic is reduced to .779 for a significance level of 85.5% indicating an almost perfect fit. This is achieved without introducing any additional parameters. The results are shown in Table 3. It can be seen that the negative binomial underestimates the frequencies in the tail indicating. Choosing the Poisson–Inverse Gaussian distribution dramatically improved the fit in the tail.

For the second data set, the negative binomial fit resulted in a significance level of .2% whereas the Poisson–Inverse Gaussian resulted in a significance level of 2.88%. The Poisson–Inverse Gaussian distribution fits much better but the fit is still marginal. When a third parameter is introduced by using the generalized Poisson-Pascal, the resulting significance level is 21.22% indicating a good fit. The results are presented in Table 4.

7. CONCLUDING REMARKS

When the distributions are fitted to data that are in the form of claim frequencies, parameter values of the fitted distribution are appropriate for a single risk ("driver" in our examples). We wish to make use of these distributions for portfolios of many risks. Since the Poisson, binomial and negative binomial distributions all have probability generating functions of

Table 3. *Fitting Bühlmann's Data*

Number of Accidents	Actual Distribution	Theoretical Distribution Neg. Bin.	P.–I.G.
0	103,704	103,723.61	103,710.03
1	14,075	13,989.95	14,054.65
2	1,766	1,857.08	1,784.91
3	255	245.19	254.49
4	45	32.29	40.42
5	6	4.24	6.94
6	2	0.56	1.26
7+	0	0.08	0.30

Table 4. *Fitting the Data of Hossack et al.*

Number of Accidents	Actual Distribution	Theoretical Distribution N.B.	P.–I.G.	G.P.P.
0	565,664	565,708.15	565,712.41	565,661.24
1	68,714	68,570.02	68,575.64	68,721.23
2	5,177	5,317.23	5,295.92	5,171.68
3	365	334.93	343.95	362.86
4	24	18.66	20.77	29.61
5	6	0.96	1.22	2.98
6+	0	0.05	0.09	0.40

the form

$$P_1(z;\theta) = [Q(z)]^\theta, \tag{17}$$

the class of contagious distributions of Section 4 have pgf's of the form

$$P(z;\theta) = [Q(P_2(z)]^\theta = [R(z)]^\theta.$$

Consequently the class is said to be "closed under convolution" since

$$[P(z;\theta)]^n = [R(z)]^{n\theta} = P(z;n\theta).$$

This means that the claim frequency distribution for a single risk remains of the same form for a portfolio of n independent risks and the computation of total claim cost distribution for portfolios of any size can be carried out using no more than two applications of the recursive formulae (3) or (5).

From equation (17), θ is a parameter of the number of accidents distribution using the interpretation described earlier. For each of the Poisson, binomial and negative binomial distributions, replacing θ by $n\theta$ results in the mean and variance of the number of accidents distribution being multiplied by a factor n. The distribution of number of claims per accident remains unaffected as one would expect.

In conclusion, the class of distributions described in this paper has a simple and logical physical interpretation, is easy to work with due to recursions (3) and (5) and fits observed data very well.

REFERENCES

Bühlmann, H. (1970), *Mathematical Methods in Risk Theory.* New York: Springer.

Douglas, J. B. (1980), *Analysis with Standard Contagious Distributions.* Fairland, Maryland: International Co-operative Publishing House.

Hogg, R. V., and S. A. Klugman (1984), *Loss Distributions.* New York: Wiley and Sons.

Hossack, I. B., J. H. Pollard, and B. Zehnwirth (1983). *Introductory Statistics with Applications in General Insurance.* Cambridge: Cambridge University Press.

Johnson, N. L., and S. Kotz (1969), *Discrete Distributions.* New York: Wiley and Sons.

Panjer, H. H. (1981), "Recursive evaluation of a family of compound distributions". *ASTIN Bulletin* **12**, 22–26.

Panjer, H. H., and G. E. Willmot (1984), "Computational techniques in reinsurance models". *Transactions of the 22nd Congress of Actuaries* **4**, 111–120.

Sundt, B., and W. Jewell (1981), "Further results on recursive evaluation of compound distributions". *ASTIN Bulletin* **12**, 27–39.

Weber, D. C. (1970), "A stochastic approach to automobile compensation". *Proceedings of the Casualty Actuarial Society* **57**, 27–63.

Willmot, G. E. (1986), "Mixed compound Poisson distributions". *ASTIN Bulletin* **17**, to appear.

R. M. Dummer [1]

ANALYZING CASUALTY INSURANCE CLAIM COUNTS

1. INTRODUCTION

When a property or casualty insurer accepts a payment of premium in return for exposure to loss, the insurer must establish a liability account, called a loss reserve, which estimates the amount of the expected loss. Accurate estimation of the expected loss is important both for the preparation of realistic financial statements and for the establishment of profitable premium rates. However, the initial loss reserve is only the beginning of the estimation problem. As the time since the period of exposure increases, the probability of a claim and hence the expectation of loss diminishes. At the same time, if a claim is made, the particular details of the claim can be used to produce a more precise estimate of loss called a case reserve. As the claim is processed, the case reserve is adjusted for payments made and new information obtained. Eventually, both types of loss reserves are eliminated and the ultimate realized loss is the total of all known payments. The estimates of claims rates and loss severities must be based on prior loss information. Because of reporting and processing delays, the most recent and hence most relevant of this information will be incomplete. Thus, the real challenge in the analysis of casualty insurance loss information is to develop methods of obtaining estimates of ultimate loss that are conditional on the current structurally incomplete loss data base.

One of the most important functions of the casualty actuary is the maintenance of informative loss reserves. The actuarial methods which have been developed for this purpose do utilize the incomplete data structure. However, the assumptions underlying these methods are extremely restrictive, the estimators tend to be rather *ad hoc*, and because there has been no attempt to develop probability models, error analyses and tests of model adequacy are completely lacking. While the literature on risk theory is rich with probability models of the loss process, these models are always param-

[1] c/o Zambia HIRD Project, Roy Littlejohn Associates, 1120 G Street, N.W., Suite 801, Washington, D.C. 20005

I. B. MacNeill and G. J. Umphrey (eds.), Actuarial Science, 127–143.

eterized in terms of final ultimate loss. In fact, applications of risk theory will commonly use the conditional actuarial estimate as the realization of ultimate loss. Practices such as this point to a definite need for statistical models which clearly recognize the realities of casualty insurance operations and address the needs of the practising actuary.

In this paper, we confine our attention solely to the problem of delays in claim reporting. Unreported losses may appear to be the most uncertain and hence most critical part of the process. In fact for most lines of business, reporting lags are relatively short. However, for liability coverages, claims may not be reported for several years and the later reported claims tend to be the most severe. Therefore, it is not surprising that the first statistical model of late reported claims was proposed by Weissner (1978) in a reinsurance context. Weissner assumed that claims are generated according to a Poisson distribution, that the underlying accident rate or equivalently claims rate per exposure unit is constant, and that the distribution of the time from loss occurrence to reporting is exponential. We extend Weissner's model by allowing the claims rate to depend on the time of the accident and by adopting a non-parametric distribution for reporting delay. This extended model is equivalent to a two factor log-linear model with no interaction (Bishop *et al.*, 1975). The maximum likelihood estimates of ultimate claim rates are identical to those produced by a development factor technique commonly employed by actuaries.[2] This relationship is extremely valuable in bridging the gap between actuarial practice and statistical modelling, highlighting the implicit assumptions of the development factor procedure, and illustrating the limitations of intuitive estimation.

In Section 2, the construction of the data base from fundamental claims processing information is described. Section 3 outlines the actuarial procedure for analyzing this data set. In Section 4, the statistical model is defined, maximum likelihood estimates derived, and the equivalence to the actuarial procedure shown. Methods of handling special situations through the use of parametric models and covariates are introduced in Section 5. The implications of these results for both actuarial practice and statistical research are discussed in the final section.

[2] Dummer and Nolan (1982) showed that for triangles of paid or incurred loss, estimates of generalized development factors could be obtained from weighted least squares regression models with some restrictions on the regression parameters and that particular choices of weights and restrictions would produce any of the most common actuarial estimates of development factors.

2. THE DATA BASE

The fundamental data base used by the practising actuary for claims analysis is a two-dimensional matrix of counts. Each claim is cross-classified by the period of exposure and the period in which the claim was reported. The most commonly applied definition of exposure period is accident period, the year or quarter in which the loss occurred. Occasionally policy period, the year or quarter in which the policy became effective, is used. We will use the generic term exposure period. The report period is the year or quarter in which the claim was reported to the insurer. However, since one of the primary variables under consideration is the time lag from loss occurrence to claim reporting, the report periods are indexed as the number of periods since the beginning of the exposure period rather than by calendar period. Thus, a loss occurring in 1982 and reported in 1984 is classified as report year three, whereas a loss also reported in 1984 but occurring in 1983 is classified as report year two. Since 1984 is the latest complete report year, no claims from accident year 1982 with a report lag greater than three could have been observed, and in this sense, the distribution of claim report lags for the 1982 accident year is truncated at three. Because the truncation point for each exposure period is determined by the date that the exposure was initiated, the resulting claim count matrix will have the familiar triangular shape. Although it is not necessary for the exposure and report periods to be of equal length, if they are equal, elements of the count matrix lying on the same diagonal will represent claims reported in the same calendar period.

The difficulty with this data structure is that a claim in the third report year does not necessarily have a report lag of between two and three years. It is the sum of time from the beginning of the exposure period to loss occurrence plus time from loss occurrence to reporting that is two to three years. The actual report lag may vary from one to three years. Similarly, true report lags for claims in the fourth report year could vary between two and four years. Hence, the observed claim count classification can distort the true underlying report lag distribution. If policy year is used as the exposure period definition, the potential distortion is even more severe.

An obvious solution to this problem is simply to use the actual report lag, which is available from claims records, to classify claims into report periods. However, the resulting data structure also has serious disadvantages. This report period classification allows the possibility that a claim from accident year 1982 in the second report period could have been reported in either 1983 or 1984. Thus, elements on the same diagonal would no longer represent claims reported in the same calendar year, and sums along diagonals may not agree with published company totals for that calendar year. In addition, the report lag distribution no longer has the same truncation point for all claims from the same exposure period. While it is theoretically possible to construct an estimate of the untruncated report lag distribution from the observed report lags and their respective truncation points, the volume of claims involved makes the data processing requirements excessive. In time, these objections may become less important. For the present, we will use the conventional data definitions and recognize the limitations of the analysis.

The measurement of the exposure base for a particular exposure period is also not without some ambiguity. For some lines, a natural exposure unit such as man hours or car years can be used. For others, it may be necessary to use earned premium which is subject to periodic rate revisions and not always directly comparable from period to period. When the exposure period is an accident year, the exposure volume for any period is not well-defined and the calendar year equivalent is usually substituted. These considerations motivate some of the modelling options discussed in Section 5.

Keeping in mind the preceeding discussion, we define the following basic data elements:

$$r = \text{report period index,} \tag{2.1}$$

$$t = \text{exposure period index,} \tag{2.2}$$

$$R_t = \text{truncation point or last observed report period for the } t\text{th exposure period} \tag{2.3}$$

and

$$T = \text{index of the most recent exposure period.} \tag{2.4}$$

Thus, t will range from 1 to T, and in the tth exposure period, r will range from 1 to R_t.

$$EP_t = \text{exposure volume for the } t\text{th exposure period,} \tag{2.5}$$

$$X_{tr} = \text{number of claims for losses occurring in the } t\text{th exposure period and reported in the } r\text{th report period.} \tag{2.6}$$

A typical claim count matrix would have the following form:

$X_{1980,1}$	$X_{1980,2}$	$X_{1980,3}$	$X_{1980,4}$	$X_{1980,5}$
$X_{1981,1}$	$X_{1981,2}$	$X_{1981,3}$	$X_{1981,4}$	
$X_{1982,1}$	$X_{1982,2}$	$X_{1982,3}$		
$X_{1983,1}$	$X_{1983,2}$			
$X_{1984,1}$				

Bearing in mind the triangular form of X, we define the row and column sums. Let

$$X_{t.} = \text{total number of claims reported to date for the } t\text{th exposure period,} \tag{2.7}$$

$$X_{.r} = \text{total number of claims reported in the } r\text{th report period,} \tag{2.8}$$

$$X_{..} = \text{total number of claims reported to date.} \tag{2.9}$$

We also require the matrix of partial sums. Let

$$S_{tr} = \text{total number of claims for the } t\text{th exposure period reported within the first } r \text{ report periods.} \tag{2.10}$$

Thus,

$$S_{tR_t} = X_{t.}. \tag{2.11}$$

The matrix S is also triangular and the row and column sums will be denoted by $S_{t.}$ and $S_{.r}$ respectively.

The following schematic diagrams illustrate which cells of X are included in each of the preceeding sums.

	000000		000000		001000
	00000		00000		00100
X_{33}	0010	$X_{3.}$	1111	$X_{.3}$	0010
	000		000		001
	00		00		00
	0		0		0

	000000		111000		111111
	00000		11100		11111
S_{33}	1110	$S_{.3}$	1110	$X_{..}$	1111
	000		111		111
	00		00		11
	0		0		1

1 – included in the sum,
0 – excluded from the sum.

3. THE ACTUARIAL APPROACH

The standard actuarial estimate of the total number of claims which will be reported for a particular exposure period is obtained by multiplying the current claim count at R_t by a factor which represents the expected percentage increase in total claims due to losses which are currently unreported. This factor, called the age-to-ultimate factor, depends on the age of the exposure period, that is the truncation point R_t. To calculate the ultimate factors, the actuary estimates the expected percentage increase in claim count due to claims that will be reported in the next report period. This factor, called the age-to-age factor, is calculated for each report period. The age-to-ultimate factor for the tth exposure period is then calculated as the product of age-to-age factors for all report periods greater than R_t. Due to the multiplicative chaining of age-to-age factors, this method is sometimes referred to as the chain-ladder method. Since the method may also be applied to triangles of paid or incurred losses, for which the period to period changes may be due to developments in previously reported claims such as payments or case reserve adjustments, rather than merely newly reported claims, the age-to-age factors are often called development factors.

Letting a_r be the age-to-age development factor for the rth report period, it follows that the age-to-ultimate factor for the same report period is given by

$$u_r = a_r a_{r+1} a_{r+2} \cdots a_{R_1}. \tag{3.1}$$

The standard actuarial estimate of a_r is the observed percentage increase from the rth to $(r+1)$th report period for all exposure periods whose current truncation point is at least $r+1$. Let

$$a_r^* = \frac{S_{.r+1}}{S_{.r+1} - X_{.r+1}}. \tag{3.2}$$

Since no claims can be reported beyond R_1, the truncation point for the oldest exposure period, a_{R_1} must be chosen *a priori*. Usually, the choice is unity but may be slightly greater than one if the actuary believes that there are still some unreported losses. In any event, the estimate of the age-to-ultimate factor is simply the appropriate product of age-to-age factor estimates. Let

$$u_r^* = a_r^* a_{r+1}^* a_{r+2}^* \cdots a_{R_1}^*. \tag{3.3}$$

If the loss occurrence rate per exposure unit is denoted m_t, then the actuarial estimate of m_t for the tth exposure period is

$$m_t^* = \frac{u_{R_t}^* S_{tR_t}}{EP_t}. \tag{3.4}$$

There is a useful probability interpretation of the age-to-age factor. Define q_r to be the conditional proability that a claim is reported in the rth report period given that it was reported within the first r periods. The logical relative frequency estimate of q_r is total claims reported in the rth period over total claims reported within the first r periods for those exposure periods whose truncation point is at least r. Letting

$$q_r^* = \frac{X_{.r}}{S_{.r}}, \tag{3.5}$$

we have:

$$a_r^* = \frac{1}{1 - q_{r+1}^*}, \tag{3.6}$$

and

$$\frac{1}{u_r^*} = (1 - q_{r+1}^*)(1 - q_{r+2}^*)(1 - q_{r+3}^*) \cdots (1 - q_{R_1}^*). \tag{3.7}$$

Thus u_r^{*-1} is the relative frequency estimate of the probability that a claim is reported within the first r report periods.

4. THE STATISTICAL MODEL

In this section, we define probability models for the loss occurrence and claim reporting processes, construct a likelihood function for the data base defined in Section 2, derive the maximum likelihood parameter estimates of the probability models, and show the equivalence of these estimates to the chain-ladder procedure.

Losses are assumed to occur according to a Poisson process. The rate of loss occurrence for the tth exposure period is taken as m_t per exposure unit and it follows that the expected number of losses or equivalently claims equals $m_t EP_t$. There is some empirical evidence which suggests that the claim generation process may not be Poisson and that a negative binomial distribution may be more appropriate. We adhere to the Poisson assumption because it is central to the results which follow. At the cost of some algebraic complexity, the same methodology can be applied to other claim generation processes.

It is assumed that a loss occurrence remains unknown until a claim is reported. The time of reporting is determined by the sum of two continuous random variables: time from the beginning of the exposure period of loss occurrence and time from loss occurrence to claim reporting. The distribution function of the sum of these two random variates is denoted $F(r)$ and is assumed to apply to all exposure periods. The probability of a loss which has occurred being reported in the rth report period is p_r, which can be computed either as the integral of $dF(r)$ over the report period interval or from

$$p_r = F(r) - F(r-1). \tag{4.1}$$

Since the distribution of X_{tr} conditional on $X_{t.}$ is multinomial with parameter p_r, the unconditional distribution of X_{tr} is Poisson with mean $EP_t m_t p_r$. By reparameterizing the mean cell count as $e^{a_0+b_t+c_r}$, this model is seen to be equivalent to a two-factor log-linear model with no interaction.

The log likelihood function of the claim count triangle is

$$L(X) = \sum_{t,r} X_{tr}(\ln EP_t + \ln m_t + \ln p_r) - EP_t m_t p_r. \tag{4.2}$$

The first order conditions for determining the maximum likelihood estimates of m_t and p_r are

$$\hat{m}_t = \frac{X_{t.}}{EP_t \hat{F}(R_t)} \tag{4.3}$$

and

$$\hat{p}_r = \frac{X_{.r}}{\sum EP_t \hat{m}_t}, \tag{4.4}$$

where the sum in the denominator extends only over those exposure periods whose truncation point is at least r.

These conditions are satisfied by

$$\hat{F}(r) = \frac{1}{u_r^*}, \tag{4.5}$$

$$\hat{p}_r = \frac{q_r^*}{u_r^*} \tag{4.6}$$

and

$$\hat{m}_t = \frac{u_{R_t}^* X_{t.}}{EP_t}. \tag{4.7}$$

We see immediately that

$$\hat{m}_t = m_t^* \tag{4.8}$$

and the estimate of ultimate claims rate obtained from the maximum likelihood estimates of this model is identical to that calculated by the actuarial procedure.

This particular solution assumes that $F(R_1)$ is unity. If $F(R_1)$ is less than one, then multiplying $\hat{p}_r$ and dividing $\hat{m}_t$ by $F(R_1)$ will give the correct solution. All other relationships remain valid.

At the current evaluation date R_t, the conditional estimate of final loss occurrence rate for the tth exposure period ought to be formed as the sum of reported claims plus estimated unreported claims. If the conditional estimate is denoted $\hat{m}_t(R_t)$, then

$$\hat{m}_t(R_t)EP_t = X_{t.} + \hat{m}_t EP_t(1 - \hat{F}(R_t)). \tag{4.9}$$

Substituting from (4.6) and (4.8) and simplifying, we have

$$\hat{m}_t(R_t)EP_t = u^*_{R_1} X_{t.}, \tag{4.10}$$

which shows that the conditional estimate equals the unconditional estimate of mean loss occurrence rate:

$$\hat{m}_t(R_t) = \hat{m}_t = m^*_t. \tag{4.11}$$

The maximum likelihood estimate of p_r has a natural probability interpretation. If q^*_r estimates the conditional probability of a claim being reported in the rth report period given that it was reported within the first r periods, then $1 - q^*$ is the conditional probability of being reported prior to r and $1/u^*_r$ is the unconditional probability of being reported by r. Thus, it follows that

$$\hat{p}_r = \frac{q^*}{u^*} \tag{4.12}$$

is the unconditional probability of being reported in the rth period.

The covariance matrix of the parameter estimates may be estimated by the standard application of large sample likelihood theory to the matrix of second derivatives of the likelihood function. The fitted values for each cell are given by

$$E_{tr} = EP_t \hat{m}_t \hat{p}_r \tag{4.13}$$

and the deviation from estimated cell count is

$$e_{tr} = X_{tr} - E_{tr}. \tag{4.14}$$

Regression diagnostics similar to those proposed by Pregibon (1981) can be applied either to the cell deviation from the fitted value or to the cell's contribution to the likelihood. The development of these diagnostics is not entirely straightforward since the triangular data format gives particular leverage to the corner points. The fitted values for the $1, R_1$ and $T, 1$ cells will always equal the observed cell counts.

In light of the uncertainties mentioned in Section 2 surrounding the measurement of exposure volume, an important characteristic of the estimates of total claims $EP_t m_t$ and report period probabilities p_r is that they depend only on the claims experience and are independent of exposure volume. However, this characteristic is a consequence of the particular model we have chosen. If the rate of claim generation is assumed to be the same for all exposure periods, then the maximum likelihood estimate of the common rate parameter m_0 is

$$\hat{m}_0 = \sum_r \frac{X_{\cdot r}}{\sum_{R_t \geq r} EP_t}. \tag{4.15}$$

The estimates of report lag probabilities are

$$\hat{p}_r = \frac{X_{\cdot r}}{\hat{m}_0 \sum_{R_t \geq r} EP_t}. \tag{4.16}$$

The sum $\sum EP_t$ in both (4.15) and (4.16) includes only those exposure periods whose truncation point is at least r. The parameter estimates, $\hat{m}_0$ and $\hat{p}_r$, are clearly dependent on EP_t. Since it is possible to select values of EP_t which will make the estimates of m_t in the basic model equal, the effect of EP_t on the parameter estimates of this model is not surprising.

The independence of $\hat{p}_r$ from exposure volume can be preserved by employing a two-stage estimate procedure. The report lag probabilities can be estimated from the full two-factor model and used to obtain a conditional maximum likelihood estimate of m_0. This procedure would be equivalent to a common actuarial practice of using the chain-ladder method to obtain estimates of ultimate counts for each exposure period and then applying either statistical or actuarial models to these estimates. This approach is applicable to the extended models discussed in Section 5.

5. EXTENSIONS OF THE BASIC MODEL

The basic model of Section 4 treated the two fundamental aspects of the claim process, loss occurrence and the claim reporting, in a very rudimentary fashion. The assumption that the report lag distribution is the same for all exposure periods is highly restrictive and probably unrealistic. Automation of claims processing procedures as well as social, political, and economic conditions will all affect the rate of claim reporting. On the other hand, assuming that each exposure period has a unique rate of loss occurrence is not restrictive enough and creates very unstable parameter estimates, particularly for the most recent exposure periods. It is reasonable to assume that adjacent exposure periods will have similiar claims experience. At the same time, claim generation rates are as dependent on external conditions as are report lags. In this section, methods are introduced for extending the basic model to address these considerations.

5.1 Models for the Report Lag Distribution

Because the report lag data for most exposure periods is truncated prior to maturity, report lag distributions which depend on the time of exposure cannot be estimated solely from the claim count data of the individual exposure period. Methods of dealing with the truncation problem have been developed for the analysis of survival data (Kalbfleisch and Prentice, 1980). The proportional hazards model incorporates a single parameter for the deviation of each exposure period's report lag distribution from the base distribution and can be applied to discrete data. The report lag distribution for the tth exposure period as specified by the proportional hazards model is

$$p_{tr} = (1 - F(r-1))^{B_t} - (1 - F(r))^{B_t}. \tag{5.1}$$

If the proportional hazard parameters B_t are completely unspecified, then the parameter for the most recent exposure period must be fixed at one since the fitted value already equals the observed count. For the second most recent period the parameter can be chosen to fit the two observed cells exactly. However, changes in the report lag distribution are likely to be evolutionary and time line models such as

$$B_t = a + bt \tag{5.2}$$

or

$$B_t = e^{a+bt} \tag{5.3}$$

may be appropriate. It may also be possible to model the proportional hazards parameter as a function of relevant covariates:

$$B_t = A\mathbf{Y}_t \tag{5.4}$$

or

$$B_t = e^{A\mathbf{Y}_t}, \tag{5.5}$$

where $\mathbf{Y}_t$ is a vector of relevant covariates.

Since the proportional hazards model may begin to over-parameterize smaller data sets, it may become advisable to consider parametric forms for the base line distribution function $F(r)$. A partial parametric form applied to the tail of the distribution may also be useful in reducing the variability in estimates of p_r for those cells in which the counts will be small.

5.2 Models for the Mean Loss Occurrence Rate

From the basic model of Section 4, the mean claim rate m_t is determined completely by the claims experience of that period in the sense that the estimate is proportional to total claims reported to date for that exposure period. For the most recent exposure period, not only will the claims rate estimate be the least reliable, but also it is not possible to detect unusual deviation due to the exact fit to the cell count. Models for the period to period behaviour of the claims rate can be used both as a device for detecting the underlying structure and as a method of applying information about the claims rates of more mature periods to the estimation of these rates for more recent exposures. Such models will reduce variability and provide meaningful indications of lack of fit.

If claims rates appear to be evolving smoothly over time, a time line regression model such as

$$m_t = a + bt \tag{5.6}$$

or

$$m_t = e^{a+bt} \tag{5.7}$$

may be appropriate. Since the notion of underwriting cycles is well established in the casualty insurance industry, cyclic functions might be considered as well. Modelling the claims rate as a function of relevant covariables is also possible:

$$m_t = A\mathbf{Y}_t \tag{5.8}$$

or

$$m_t = e^{A\mathbf{Y}_t}, \tag{5.9}$$

where $\mathbf{Y}_t$ is a vector of relevant covariables.

If the claims for each exposure period are segregated into rate-making groups, rating models could be developed:

$$m_{tk} = f(\mathbf{Y}_t, \mathbf{Z}_k), \tag{5.10}$$

where $\mathbf{Z}_k$ is a vector of variables relating to the kth rate-making group.

In Section 4, we noted the effect of the definition of exposure volume EP_t on parameter estimates when parametric models are used. One suggested solution to this problem is a two-stage estimation procedure. An alternative is the application of several definitions of exposure volume: natural exposure units, actual premium, and premium adjusted for rate revisions or inflation.

A common actuarial solution to the problem of excessive variability caused by limited experience is a method of empirical Bayes or Stein shrinkage estimation called credibility. The credibility estimate of mean claim rate for a particular exposure period is a weighted average of the claims rate for that exposure period with the overall claims rate for all exposure periods:

$$m_t^c = z_t\hat{m}_t + (1 - z_t)\hat{m}_0. \tag{5.11}$$

The credibility factor z_t is determined by the volume of claims experience and variation in claims rates among the exposure periods. Using standard methods (Bühlmann and Straub, 1970; Jewell, 1976) and assuming that the exposure volume for the tth exposure period is $EP_tF(R_t)$, it is possible to construct a facsimile of the credibility factor from the parameter estimates for the two basic models of Section 4. However, the estimates $\hat{m}_t$ are not mutually independent. An exact empirical Bayes solution is an open and interesting theoretical problem.

The specification of prospective models introduced in this section is intentionally vague. It is only through empirical research with actual claims experience that workable models will be developed.

6. SUMMARY AND CONCLUSIONS

We have shown that a particular actuarial technique for estimating ultimate claim counts from incomplete data produces the same results as the maximum likelihood estimates of a well-defined probability model. It is not surprising that a technique, however naive, which has survived many years of practical application is found to have some theoretical basis. A little investigation would discover similar bases for many procedures commonly used in the analysis of financial data. Explicit identification of the underlying statistical model increases the information that is available about the procedures, improves understanding of its characteristics, and leads to more effective application of its results.

By establishing a statistical model for the claim count data, we have been able to derive an explicit estimate of the report lag distribution, calculate standard errors for parameter estimates, and produce estimated cell counts which can be used to identify trends or outliers. In addition, the assumptions and limitations of the method are clearly stated. The advantages of the statistical method for the basic model is enough to justify its use. However, the basic model is rarely adequate and it is the inability of the intuitive actuarial procedure to incorporate extensions directly that is its real limitation. In the face of rapidly changing conditions, actuaries must resort to judgmental adjustments of the basic parameters. Because these adjustments are not procedural, they can never be tested or evaluated. Covariates measuring external conditions can be incorporated into the statistical model in a natural and consistent fashion. Standard statistical procedures not only produce more precise estimates of covariate effects, but also permit testing and possible rejection for hypothesized effects. In addition, utilizing the statistical approach requires a clear definition of the mechanism by which the covariates affect cell counts, and this enforced discipline has been known to have a sobering effect on many an otherwise cavalier investigator.

The equivalence between basic statistical models and actuarial procedures has the potential to be productively exploited in introducing statistical methodology into actuarial practice. However, it is important to point out that the simple closed form structure of the actuarial estimators does allow the practitioner to easily determine the effect of individual cell counts on these estimators. If more sophisticated statistical methods are to be adopted in practice, they must be accompanied by the kind of diagnostic tools which facilitate meaningful sensitivity analysis.

APPENDIX

We demonstrate that the first order maximum likelihood conditions

$$\hat{m}_t = \frac{X_{t.}}{EP_t\hat{F}(R_t)} \tag{4.3}$$

and

$$\hat{p}_r = \frac{X_{.r}}{\sum EP_t\hat{m}_t} \tag{4.4}$$

are satisfied by

$$\hat{F}(r) = \frac{1}{u_r^*}, \tag{4.5}$$

$$\hat{p}_r = \frac{q_r^*}{u_r^*} \tag{4.6}$$

and

$$\hat{m}_t = \frac{u_{R_t}^* X_{t.}}{EP_t}. \tag{4.7}$$

Equation (4.7) follows directly from (4.5) by substitution. The equivalence of Equations (4.5) and (4.6) follows from

$$p_r = F(r) - F(r-1) \tag{4.1}$$

and

$$\frac{1}{u_{r-1}^*} = \frac{(1-q_r^*)}{u_r^*} \tag{A.1}$$

from (3.7). Thus by substitution

$$\hat{p}_r = \frac{q_r^*}{u_r^*}. \tag{A.2}$$

It remains to show that (4.6) satisfies (4.4), i.e.,

$$\frac{q_r^*}{u_r^*} = \frac{X_{.r}}{\sum EP_t \hat{m}_t}. \tag{A.3}$$

The demonstration will proceed by induction. Suppose (A.3) is true for all $r > r'$. In particular,

$$\frac{q_{r'+1}^*}{u_{r'+1}^*} = \frac{X_{.r'+1}}{\sum EP_t \hat{m}_t} \tag{A.4}$$

or

$$\sum EP_t \hat{m}_t = \frac{X_{.r'+1} u_{r'+1}^*}{q_{r'+1}^*}, \tag{A.5}$$

where the sum is over all t such that $R_t > r'$. Since (A.3) is true for all $r > r'$, it follows from (A.1) that

$$\hat{F}(r') = \hat{F}(r'+1) - \hat{p}_{r+1} = \frac{1}{u_{r+1}^*} - \frac{q_{r+1}^*}{u_{r+1}^*} = \frac{1}{u_r^*} \tag{A.6}$$

and by substitution into (4.3)

$$\hat{m}_{t'} = \frac{X_{t'.} u_{r'}^*}{EP_{t'}}. \tag{A.7}$$

For any exposure period t' whose truncation point is r', that is,

$$R_{t'} = r', \tag{A.8}$$

the sum in the denominator of (4.4) can be partitioned into those exposure periods with $R_t > r'$ and those with $R_t = r'$. Therefore

$$\hat{p}_r = \frac{X_{.r'}}{\frac{X_{.r'+1} u_{r'+1}^*}{q_{r'+1}^*} + u_{r'}^* \sum X_{t'.}} \tag{A.9}$$

by substitution from (A.5). The sum in the denominator is over all exposure periods having $R_{t'} = r'$. From (A.1) we have

$$\hat{p}_{r'} = \frac{1}{u_{r'}^*} \left(\frac{X_{.r'}}{\frac{(1-q_{r'+1}^*) X_{.r'+1}}{q_{r'+1}^*} + \sum X_{t'.}} \right). \tag{A.10}$$

Now since

$$q_r^* = \frac{X_{.r}}{S_{.r}}, \tag{3.5}$$

it follows that

$$\hat{p}_{r'} = \frac{1}{u_{r'}^*}\left(\frac{X_{.r'}}{S_{.r'+1} - X_{.r'+1} + \sum X_{t'.}}\right). \tag{A.11}$$

We note that

$$S_{.r'} = S_{.r'+1} - X_{.r'+1} + \sum X_{t'.} \tag{A.12}$$

Since $S_{.r}$ is the total of all claims reported by r for all exposure periods with $R_t = r$, $X_{.r+1}$ is the total of all claims included in $S_{.r+1}$ but not $S_{.r}$, and $\sum X_{t'.}$ is the total of all claims included in $S_{.r}$ but not in $S_{.r+1}$. By substitution into (A.11)

$$\hat{p}_{r'} = \frac{1}{u_{r'}^*}\left(\frac{X_{.r'}}{S_{.r'}}\right) = \frac{q_{r'}^*}{u_{r'}^*}, \tag{A.13}$$

which proves the induction step.

Finally, since both $\hat{F}(R_1) = 1$ and $u_{R_1}^* = 1$,

$$\hat{m}_{t''} = \frac{X_{t''.}}{EP_{t''}} \tag{A.14}$$

for all exposure periods t'' having $R_{t''} = R_1$. Since no exposure periods have a truncation point exceeding R_1, by summing over t'' we have

$$\sum EP_{t''}\hat{m}_{t''} = S_{.R_1} \tag{A.15}$$

and from (4.4)

$$\hat{p}_{R_1} = \frac{X_{.R_1}}{S_{.R_1}} = q_{R_1}^*, \tag{A.16}$$

which proves the proposition is true at the initial step.

REFERENCES

Bishop, Y. M. M., S. E. Fienberg, and P. W. Holland (1975), *Discrete Multivariate Analysis: Theory and Practice*. Cambridge, MA: MIT Press.

Bühlmann, H., and E. Straub (1970), "Glaubwurdigkeit fur Schadensatze". *Bulletin de l'Association des Actuaries Suisses* **70**, 111–133. (Translation: "Credibility for Loss Ratios" by C. E. Brooks in *ARCH*, 1972.)

Dummer, R. M., and J. D. Nolan (1982), "The application of linear regression models to loss development factor estimation". *Casualty Loss Reserve Seminar, American Academy of Actuaries.*

Kalbfleisch, J. D., and R. L. Prentice (1980), *The Statistical Analysis of Failure Time Data.* New York: Wiley and Sons.

Jewell, W. S. (1976), "A survey of credibility theory". ORC Report 76–31, Operations Research Center, University of California, Berkeley.

Morris, C., and O. E. Van Slyke (1978), "Empirical Bayes methods for pricing insurance classes". *Proceedings of the Business and Economic Statistics Section, American Statistical Association.*

Pregibon, D. (1981), "Logistic regression diagnostics". *Annals of Statistics* **9**, 705–724.

Skurnick, D. (1973), "A survey of loss reserving methods". *Proceedings of the Casualty Actuarial Society* **60**, 16–62.

Weissner, E. W. (1978), "Estimation of the distribution of report lags by the method of maximum likelihood". *Proceedings of the Casualty Actuarial Society* **65**, 1–9.

Elias S. W. Shiu [1]

IMMUNIZATION—THE MATCHING OF ASSETS AND LIABILITIES

ABSTRACT

A major problem facing the insurance industry today is the matching of the asset and liability cashflows so as to minimize the risks arising from interest rate fluctuations. Immunization is a technique used by actuaries and investment professionals to tackle this problem. This paper gives a brief review of Redington's theory of immunization and discusses its extensions by the theory of inequalities of convex functions.

1. INTRODUCTION

Consider a block of insurance business and its associated assets. For $t \geq 0$, let A_t denote the *asset cashflow* expected at time t, i.e., the investment income and capital maturities expected at that time. Let L_t denote the *liability cashflow* expected at time t, i.e., the policy claims plus policy surrenders plus expenses minus premium income expected at that time. Define the *net cashflow* at time t, $t \geq 0$, as

$$N_t = A_t - L_t.$$

The theory of immunization is concerned with two kinds of risk that may occur due to interest rate fluctuations. (1) Positive net cashflows may have to be reinvested at lower interest rates. (2) Negative net cashflows may involve the liquidations of assets at depreciated values because of higher interest rates.

[1] Department of Actuarial and Management Sciences, University of Manitoba, Winnipeg, Manitoba, R3T 2N2; also The Great-West Life Assurance Company, Winnipeg, Manitoba R3C 3A5

I. B. MacNeill and G. J. Umphrey (eds.), Actuarial Science, 145–156.

2. REDINGTON'S THEORY OF IMMUNIZATION

Given a force of interest δ, the surplus of the block of business is the present value of the net cashflows evaluated at δ,

$$S(\delta) = \sum_{t \geq 0} N_t e^{-\delta t}.$$

If the force of interest changes from δ to $\delta + \epsilon$, the surplus value changes from $S(\delta)$ to $S(\delta + \epsilon)$. What are the conditions on the cashflows such that

$$S(\delta + \epsilon) \geq S(\delta)?$$

By Taylor's expansion,

$$\begin{aligned} S(\delta + \epsilon) - S(\delta) &= \sum_{t \geq 0} N_t e^{-\delta t} \left(e^{-\epsilon t} - 1\right) \\ &= \sum_{t \geq 0} N_t e^{-\delta t} \left(-\epsilon t + \epsilon^2 t^2/2 - \cdots\right). \end{aligned}$$

Thus, if

$$\sum t N_t e^{-\delta t} = 0 \tag{2.1}$$

and

$$\sum t^2 N_t e^{-\delta t} > 0, \tag{2.2}$$

we have

$$S(\delta + \epsilon) > S(\delta)$$

for $|\ \epsilon\ |$ sufficiently small.

3. A GENERALIZATION OF REDINGTON'S THEORY

It has been pointed out by Fisher and Weil (1971, p. 417) that Redington's model should be generalized to the case where both the force of interest δ and the shock ϵ are functions of time. Thus, we consider

$$S(\delta) = \sum_{t \geq 0} N_t \exp\left(-\int_0^t \delta(s) ds\right)$$

and

$$S(\delta + \epsilon) = \sum_{t \geq 0} N_t \exp\left(-\int_0^t (\delta(s) + \epsilon(s)) ds\right).$$

In the finance literature, the force-of-interest function $\delta(\cdot)$ is called the *instantaneous forward-rate* function. Define

$$V_t = N_t \exp\left(-\int_0^t \delta(s)ds\right)$$

and

$$f(t) = \exp\left(-\int_0^t \epsilon(s)ds\right).$$

Then

$$S(\delta+\epsilon) - S(\delta) = \sum_{t\geq 0} V_t(f(t) - 1).$$

By Taylor's formula with remainder,

$$\begin{aligned} f(t) &= f(0) + tf'(0) + \int_0^t (t-w)f''(w)dw \\ &= 1 - t\epsilon(0) + \int_0^t (t-w)f''(w)dw. \end{aligned}$$

Thus,

$$S(\delta+\epsilon) - S(\delta) = -\epsilon(0)\sum tV_t + \sum V_t \int_0^t (t-w)f''(w)dw.$$

Suppose that the cashflows can be structured such that

$$\sum tV_t = 0;$$

then

$$S(\delta+\epsilon) - S(\delta) = \sum_{t\geq 0} V_t \int_0^t (t-w)f''(w)dw.$$

With the definition

$$(t-w)_+ = \max\{0, t-w\},$$

the last equation becomes

$$\begin{aligned} S(\delta+\epsilon) - S(\delta) &= \sum_{t\geq 0} V_t \int_0^\infty (t-w)_+ f''(w)dw \\ &= \int_0^\infty (f''(w)\sum_{t\geq 0} V_t(t-w)_+)dw. \end{aligned}$$

Hence, a sufficient condition for the inequality

$$S(\delta+\epsilon) \geq S(\delta) \tag{3.1}$$

to hold is that

$$\sum tV_t = 0$$

and

$$f''(w)\sum V_t(t-w)_+ \geq 0, \qquad \text{for all } w \geq 0.$$

Consequently, if it is forecasted that

$$f''(w) \geq 0, \qquad \text{for all } w \geq 0, \tag{3.2}$$

it would be advantageous to arrange the cashflows such that

$$\sum tV_t = 0 \tag{3.3}$$

and

$$\sum V_t(t-w)_+ \geq 0, \qquad \text{for all } w \geq 0. \tag{3.4}$$

Since

$$f''(w) = f(w)\left[(\epsilon(w))^2 - \epsilon'(w)\right], \tag{3.5}$$

we have

$$f''(w) \geq 0$$

if and only if

$$(\epsilon(w))^2 \geq \epsilon'(w). \tag{3.6}$$

Note that inequality (3.6) holds for all w if $\epsilon(w)$ is a nonincreasing function of w. Hence, if $\epsilon(w)$ is a constant function, (3.2) is satisfied. Thus, relations (3.3) and (3.4) can be viewed as a generalization of Redington's conditions (2.1) and (2.2). Moreover, if it is predicted that the next interest rate shock will satisfy the reverse of inequality (3.6) for each w, to obtain (3.1) one would try to arrange the cashflows to satisfy (3.3) and the reverse of inequalities (3.4).

The theory above can be further generalized using a few more terms of the Taylor expansion:

$$f(t) = f(0) + tf'(0) + t^2 f''(0)/2 + \cdots + t^n f^{(n)}(0)/n! + \left[\int_0^t (t-w)^n f^{(n+1)}(w)\,dw\right]/n!.$$

Indeed, the results in Section II of Chan (1985) can be derived with this formula. Although $f^{(n+1)}(w)$ can be obtained using the Faà di Bruno formula, for $n > 1$ the equation corresponding to (3.5) is much more complicated. Thus, for $n > 1$, we are not able to characterize

$$f^{(n+1)}(w) \geq 0$$

in terms of a simple inequality such as (3.6). We would also like to remark that, if

$$\epsilon(w) \neq 0,$$

inequality (3.6) is equivalent to

$$-1 \leq \left(1/\epsilon(w)\right)'.$$

Now, suppose that the values of t are positive integers only, i.e., the cashflows always occur at year-ends. Then, instead of Taylor's formula, we can use Newton's forward-difference formula:

$$f(t) = f(0) + \binom{t}{1}\Delta f(0) + \binom{t}{2}\Delta^2 f(0) + \cdots + \binom{t}{n}\Delta^n f(0)$$
$$+ \sum_{j \geq 1} \binom{t-j}{n} \Delta^{n+1} f(j-1).$$

A proof of this formula can be found on page 226 of Goovaerts *et al.* (1984). Consequently, if it is believed that

$$\Delta^2 f(z) \geq 0, \quad \text{for } z = 0, 1, 2, \ldots,$$

one would attempt to structure the cashflows so that (3.3) holds and

$$\sum V_t (t-k)_+ \geq 0, \qquad k = 1, 2, 3, \ldots. \tag{3.7}$$

Note that it is much simpler to deal with (3.7) than with (3.4). Also observe that for each x, given a positive integer m, there exists by the mean value theorem a value ς between x and $x + m$ such that

$$\Delta^m f(x) = f^{(m)}(\varsigma)/m!.$$

Hence, the condition

$$f^{(m)}(w) \geq 0, \qquad \text{for all } w \geq 0,$$

implies that

$$\Delta^m f(v) \geq 0, \qquad \text{for all } v \geq 0.$$

4. EQUIVALENT CONDITIONS

For the rest of this section we assume that all cashflows occur at year-ends; thus,

$$\{V_t\} = \{V_1, V_2, V_3, \ldots,\}$$

and condition (3.7) replaces (3.4). We also assume that (3.3) holds and the next interest rate shock $\epsilon(\cdot)$ will satisfy (3.6). It follows from (3.3) and (3.7) that

$$0 \geq \sum_{t\geq 1} (t - (t-k)_+)V_t, \qquad k = 1, 2, 3, \ldots.$$

For $k = 1$, we have

$$\begin{aligned} 0 &\geq \sum_{t\geq 1} V_t \\ &= S(\delta). \end{aligned}$$

This is indeed a surprising constraint, since it means that there is no general way to protect a *positive* surplus against all interest rate shocks satisfying (3.6). For the rest of this paper, we shall assume that

$$S(\delta) = \sum_{t\geq 1} V_t = 0. \tag{4.1}$$

We now give two conditions which are equivalent to (3.7) when both (4.1) and (3.3) hold. Since

$$(t-k)_+ = (\mid t-k \mid + t - k)/2$$

and

$$\begin{aligned} \sum (t-k)V_t &= \sum tV_t - k\sum V_t \\ &= 0 - k \cdot 0 \\ &= 0, \end{aligned}$$

we have

$$\sum (t-k)_+ V_t \geq 0$$

if and only if

$$\sum \mid t-k \mid V_t \geq 0,$$

i.e., if and only if

$$\sum \mid t-k \mid A_t \exp\left(-\int_0^t \delta(s)ds\right) \geq \sum \mid t-k \mid L_t \exp\left(-\int_0^t \delta(s)ds\right).$$

The last inequality has been derived by Fong and Vasicek (1983a; 1983b, p. 232) and they call it the mean absolute deviation (MAD) constraint. It implies that the asset cashflows should be more dispersed than the liability cashflows.

The second condition equivalent to (3.7) is a classical result in mathematical analysis due to J. Karamata (1932). (Also see Marshall and Olkin (1979, p. 449) and Karlin and Studden (1966, p. 421, p. 425).) Since

$$(t-k)_+ = (k-t)_+ + t - k,$$

by (4.1) and (3.3) we have

$$\begin{aligned}\sum_{t\geq 1}(t-k)_+ V_t &= \sum_{t\geq 1}(k-t)_+ V_t \\ &= \sum_{k-1\geq j\geq 1}\left(\sum_{j\geq t\geq 1} V_t\right).\end{aligned}$$

Thus

$$\sum_{t\geq 1}(t-k)_+ V_t \geq 0$$

if and only if

$$\sum_{k-1\geq j\geq 1}\left(\sum_{j\geq t\geq 1} V_t\right) \geq 0.$$

Karamata's result is actually more general, as follows.

Theorem 4.1. Let μ be a signed measure defined on the Borel subsets of (a, b). Then

$$\int_a^b \varphi d\mu \geq 0, \qquad \text{for all convex functions } \varphi,$$

if and only if

$$\int_a^b 1 d\mu = \int_a^b x d\mu = 0$$

and

$$\int_a^t \mu(a, x]dx \geq 0, \qquad a \leq t \leq b.$$

We can use Karamata's Theorem to treat the case where the cashflows do not necessarily occur at year-ends, i.e., the time intervals between cashflows are irregular. Karamata's Theorem is related to the following celebrated result of Hardy, Littlewood and Pólya (1929, 1952).

Theorem 4.2. Given two *ordered* sequences of real numbers

$$x_1 \geq x_2 \geq \cdots \geq x_n$$

and

$$y_1 \geq y_2 \geq \cdots \geq y_n,$$

the following three conditions are equivalent.

(i) $\sum_{1 \leq i \leq n} \varphi(x_i) \leq \sum_{1 \leq j \leq n} \varphi(y_j)$ for all continuous convex functions φ.

(ii) $\sum_{1 \leq i \leq k} x_i \leq \sum_{1 \leq j \leq k} y_j, \quad k = 1, 2, \ldots, n-1$, and $\sum_{1 \leq i \leq n} x_i = \sum_{1 \leq j \leq n} y_j$.

(iii) There exists a doubly stochastic matrix A such that $(x_1, x_2, \ldots, x_n) = (y_1, y_2, \ldots, y_n)A$.

One may wonder what part (iii) of Theorem 4.2 means in the context of immunization. The following result, due to David Blackwell (1951, 1953) (also see Marshall and Olkin, 1979, Chapter 14; Kemperman, 1975, p. 114; Blackwell and Gerschick, 1954, Section 12.2), generalizes the equivalence between (i) and (iii) of Theorem 4.2.

Theorem 4.3. Let $\mathbf{x}, \mathbf{p} \in \mathcal{R}^m$ and $\mathbf{y}, \mathbf{q} \in \mathcal{R}^n$. Assume that

$$\sum_{1 \leq i \leq m} p_i = \sum_{1 \leq j \leq n} q_j = 1$$

and

$$p_i, q_j \geq 0.$$

Then

$$\sum_{1 \leq i \leq m} p_i \varphi(x_i) \leq \sum_{1 \leq j \leq n} q_j \varphi(y_j)$$

for all continuous convex functions φ if and only if there exists a nonnegative $n \times m$ matrix B with its column sums all equal to 1 such that

$$B\mathbf{p} = \mathbf{q}$$

and

$$\mathbf{x}^T = \mathbf{y}^T B.$$

In terms of immunization theory, Theorem 4.3 is not quite what we are looking for. The desired result also follows from Blackwell's work, although it is stated more clearly by Sherman (1951).

Theorem 4.4. Let $\mathbf{x}, \mathbf{p} \in \mathcal{R}^m$ and $\mathbf{y}, \mathbf{q} \in \mathcal{R}^n$. Assume that

$$\sum_{1 \le i \le m} p_i = \sum_{1 \le j \le n} q_j$$

and

$$p_i, q_j \ge 0.$$

Then

$$\sum_{1 \le i \le m} p_i \varphi(x_i) \le \sum_{1 \le j \le n} q_j \varphi(y_j)$$

for all continuous convex functions φ if and only if there exists a nonnegative $m \times n$ matrix C with all its column sums equal to 1 such that

$$\mathbf{p} = C\mathbf{q}$$

and

$$(x_1 p_1, x_2 p_2, \ldots, x_m p_m)^T = C\,(y_1 q_1, y_2 q_2, \ldots, y_n q_n)^T.$$

Now, consider p_i as the present value of the ith liability cashflow which is expected to occur at time x_i, and q_j, the present value of the jth asset cashflow expected to occur at time y_j. To apply Theorem 4.4 we assume that none of the asset and liability cashflows are negative numbers. The second half of the theorem says that there exists a *matching* between the asset and liability cashflows—there is a partition of the asset cashflows into m streams, each with the same present value and duration as one of the m liability cashflows. Thus, in a sense, the problem of multiple liability immunization is reduced to the problem of single liability immunization.

Theorem 4.4 has also been proved by Fong and Vasicek (1983a). The second half of this interesting paper is not yet published. The first half deals with the single liability case and has appeared as Fong and Vasicek (1984).

Let me conclude this section with two remarks. (1) There is another condition equivalent to each of the three conditions in Theorem 4.2. It is a result due to Horn (1954) and Mirsky (1958): there exists a real symmetric matrix with diagonal entries $x_1, x_2, \ldots, x_n$ and eigenvalues $y_1, y_2, \ldots, y_n$. However, I am not able to reformulate this condition into something useful to immunization theory. (2) For the reader interested in the generalization of the Hardy, Littlewood and Pólya Theorem in function spaces, I would recommend Luxemburg (1967).

5. EXACT MATCHING

Assume that $\sum V_t = 0$ and $\sum tV_t = 0$. Thus, for a twice-differentiable function φ,

$$\sum_{t\geq 0} V_t\varphi(t) = \int_0^\infty \left(\varphi''(w)\sum_{t\geq 0} V_t(t-w)_+\right) dw. \tag{5.1}$$

Now, also assume (3.4) to hold. If φ is convex, by (5.1)

$$\sum_{t\geq 0} V_t\varphi(t) \geq 0.$$

If φ is strictly convex ($\varphi'' > 0$), then

$$\sum_{t\geq 0} V_t\varphi(t) = 0$$

if and only if

$$\sum_{t\geq 0} V_t(t-w)_+ = 0, \qquad \text{for all } w \geq 0,$$

which means that, for all t,

$$V_t = 0$$

or

$$A_t = L_t,$$

i.e., exact cashflow matching. Thus, if the cashflows are expected to occur at year-ends, we can get close to exact cashflow matching as follows. Choose an appropriate strictly convex function φ. Using the method of linear programming, minimize

$$\sum_{t\geq 1} V_t\varphi(t)$$

subject to (4.1), (3.3) and (3.7). Fong and Vasicek (1983a,b) have advocated using the function

$$\varphi(t) = (t - D)^2,$$

where

$$D = \left(\sum L_t \exp(-\int_0^t \delta(s)ds)\right)^{-1} \sum tL_t \exp\left(-\int_0^t \delta(s)ds\right)$$
$$= \left(\sum A_t \exp(-\int_0^t \delta(s)ds)\right)^{-1} \sum tA_t \exp\left(-\int_0^t \delta(s)ds\right).$$

However, since (4.1) and (3.3) hold, for each z

$$\sum V_t(t-D)^2 = \sum V_t(t-z)^2.$$

In Section 3 of Shiu (1986) it is shown that, subject to (3.3) and (3.4), there exists a positive number ξ, which depends on ϵ and V_t, such that

$$S(\epsilon + \delta) - S(\delta) = \frac{1}{2} f''(\xi) \sum V_t \cdot t^2.$$

Thus, an appropriate strictly convex function is

$$\varphi(t) = t^2.$$

REFERENCES

Blackwell, D. (1951), "Comparison of experiments." *Proceedings of the Second Berkeley Symposium on Mathematical Statistics and Probability*, pp. 93–102.

Blackwell, D. (1953), "Equivalent comparisons of experiments." *Annals of Mathematical Statistics* **24**, 265–272.

Blackwell, D., and M. A. Gerschick (1954), *Theory of Games and Statistical Decisions*. New York: Wiley. Reprinted by Dover.

Chan, B. (1985), "Stochastic ordering characterization of immunization." Working Paper TR–85–18, Department of Statistical and Actuarial Sciences, The University of Western Ontario.

Fisher, L., and R. L. Weil (1971), "Coping with the risk of interest-rate fluctuations: Returns to bondholders from naive and optimal strategies." *Journal of Business* **44**, 408–431.

Fong, H. G., and O. Vasicek (1983a), "A risk minimizing strategy for multiple liability immunization." Unpublished.

Fong, H. G., and O. Vasicek (1983b), "Return maximization for immunized portfolios." In *Innovations in Bond Portfolio Management: Duration Analysis and*

Immunization, ed. G. G. Kaufman, G. O. Bierwag and A. Toevs, pp. 227–238. Greenwich, Connecticut: JAI Press.

Fong, H. G., and O. Vasicek (1984), "A risk minimizing strategy for portfolio immunization." *Journal of Finance* **39**, 1541–1546.

Goovaerts, M. J., F. de Vylder, and J. Haezendonck (1984), *Insurance Premiums*. Amsterdam: North-Holland.

Hardy, G. H., J. E. Littlewood, and G. Pólya (1929), "Some simple inequalities satisfied by convex functions." *Messenger of Mathematics* **58**, 145–152.

Hardy, G. H., J. E. Littlewood, and G. Pólya (1952), *Inequalities*, 2nd edition. London: Cambridge University Press.

Horn, A. (1954), "Doubly stochastic matrices and the diagonal of a rotation matrix." *American Journal of Mathematics* **76**, 620–630.

Karamata, J. (1932), "Sur une inégalité relative aux fonctions convexes." *Publications Mathématiques de l'Université Belgrade* **1**, 145–148.

Karlin, S. J., and W. J. Studden (1966), *Tchebycheff System: With Applications in Analysis and Statistics*. New York: Wiley.

Kemperman, J. H. B. (1975), "The dual of the cone of all convex functions on a vector space." *Aequationes Mathematicae* **13**, 103–119.

Luxemburg, W. A. J. (1967), "Rearrangement-invariant Banach function spaces." In *Proceedings of the Symposium in Analysis, Queen's Papers in Pure and Applied Mathematics—No. 10*, Queen's University, Kingston, Ontario, pp. 83–144.

Marshall, A. W., and I. Olkin (1979), *Inequalities: Theory of Majorization and Its Applications*. New York: Academic Press.

Mirsky, L. (1958), "Matrices with prescribed characteristic roots and diagonal elements." *Journal of the London Mathematical Society* **33**, 14–21.

Redington, F. M. (1952), "Review of the principles of life-office valuations." *Journal of the Institute of Actuaries* **78**, 286–315.

Schmeidler, D. (1979), "A bibliographical note on a theorem of Hardy, Littlewood, and Pólya." *Journal of Economic Theory* **20**. 125–128.

Sherman, S. (1951), "On a theorem of Hardy, Littlewood, Pólya, and Blackwell." *Proceedings of the National Academy of Sciences of the United States of America* **37**, 826–831.

Shiu, E. S. W. (1986), "A generalization of Redington's theory of immunization." Actuarial Research Clearing House. To appear.

John A. Beekman [1]

ORNSTEIN-UHLENBECK STOCHASTIC PROCESSES APPLIED TO IMMUNIZATION

ABSTRACT

The purposes of this paper are to describe some results about the term structure of interest rates as stochastic processes, and to extend to this topic some of the author's earlier results on stochastic processes. It is assumed that the instantaneous borrowing and lending rate (the spot rate) is modelled by a stochastic process which is both Markovian and Gaussian. Among such processes, the Ornstein-Uhlenbeck stochastic process is used. It has the advantage of being able to model phenomena which react to offset excessive movements in any one direction. Probabilities that the spot rates will deviate from the long term mean by more than preassigned boundary functions during various time intervals are obtained. The boundary functions are of the form $A\sigma$ or $A\sigma(1+R)^t$, $0 \leq t \leq T$. Knowledge of such probabilities is needed for some applications of immunization theory.

1. INTRODUCTION

Let $P(t,s)$ denote the price at time t of a default free discount bond (i.e., a zero coupon bond) maturing at time s, $t \leq s$, with unit maturity value, $P(s,s) = 1$. The yield to maturity $R(t,T)$ is the internal rate of return at time t on a bond with maturity date $s = t+T$. The collection of rates $\{R(t,T), t \leq T \leq s\}$ is called the term structure at time t. The spot rate is defined as the instantaneous borrowing and lending rate,

$$r(t) = R(t,0) = \lim_{T \to 0} R(t,T).$$

[1] Department of Mathematical Sciences, Ball State University, Muncie, Indiana 47306

I. B. MacNeill and G. J. Umphrey (eds.), Actuarial Science, 157–164.

Vasicek (1977) assumed that the spot rate follows a continuous Markov process. It is quite reasonable to assume the stochastic process has continuous sample paths because $r(t)$ does not change value by an instantaneous jump. The Markovian assumption means the future development of the spot rate given its present value is independent of the past development that has led to the present level. It will also be assumed that the process is Gaussian, which means that any finite collection of random variables $\{r(t_i),\ i = 1, 2, \cdots, n\}$ has a multivariate normal distribution.

The most widely studied stochastic process which is both Gaussian and Markovian is the Wiener process $\{w(t),\ 0 \leq t < \infty\}$ with mean function $E\{w(t)\} \equiv 0$ for all t and covariance function $E\{w(s)w(t)\} =$ minimum (s, t), for $0 \leq s,\ t < \infty$. Among its many purposes is its use in describing diffusion processes (those which are Markovian and have continuous sample paths). Thus the $r(t)$ process can be expressed through the stochastic differential equation

$$dr = f(r, t)dt + \rho(r, t)dw,$$

where $w(t)$ is a Wiener process with incremental variance dt. The functions $f(r, t)$and $\rho^2(r, t)$ are the instantaneous drift and the variance of the spot rate process.

In addition to Vasicek (1977), Cox *et al.* (1981), Brennan and Schwartz (1977), Richard (1978), and Boyle (1978, 1980) have modelled the $r(t)$ process by a Gaussian Markov stochastic process. The concept of immunization within the framework of a stochastic model for $r(t)$ is examined by Boyle (1978, 1980). The present author has used such processes to model random deviations fom actuarial assumptions about investment performance, operating expenses, and lapse expenses (Beekman, 1974, 1975; Beekman and Fuelling, 1977, 1979, 1980). In this paper, the time variables s and t will be reversed from their uses in the earlier papers.

Vasicek's development leads to the more specific equation

$$dr = \alpha(\gamma - r)dt + \rho dw,$$

where α, γ, and ρ are positive constants. This corresponds to the Ornstein-Uhlenbeck process, the second most studied and used stochastic process which is both Gaussian and Markovian.

2. THE ORNSTEIN-UHLENBECK PROCESS

This process will now be described in general terms because there are differences in how the cited papers use that random process. The Ornstein-Uhlenbeck (O.U.) process was developed by research physicists to model the velocities of Brownian motion. It has a constant variance function σ^2. This is in marked contrast to the unbounded variance function for the Wiener process, i.e., $\text{Var}\{w(t)\} = t$, $0 \leq t < \infty$. Moreover, the O.U. process has the advantage of being able to model phenomena which react to offset excessive movements in any one direction. Thus, if $\{X(t),\ 0 \leq t < \infty\}$ is an O.U. process, the conditional mean function $E\{X(s) \mid X(t) = x\} = xe^{-\alpha(s-t)}$ for $\alpha > 0$ and $t < s$. This implies a drift downward if the present deviation is positive and a drift upward if the present deviation is negative. By contrast, the conditional mean function for the Wiener process is $E\{w(s) \mid w(t) = x\} = x$, which does not reflect any stabilizing tendency. Following an equation similar to

$$dr = \alpha(\gamma - r)dt + \rho dw,$$

Cox *et al.* (1981, p. 790) state, "This model of interest rate dynamics dates back, at least in spirit, to the Keynesian (1930) concept of a "return to a normal level" of interest rates, and is characterized by the following behavior. Over the next period, the interest rate when above (below) its usual level is expected to fall (rise) by an amount proportional to the current deviation from normal."

Let us assume sample information is available on the process. Denote the values by X_i, $i = 1, 2, \cdots, n$, with relative frequencies $f(i)$, $i = 1, 2, \cdots, n$. Then

$$\sum_{i=1}^{n} X_i^2 f(i) - [\sum_{i=1}^{n} X_i f(i)]^2$$

could be used to estimate σ^2.

For an O.U. process, the covariance function

$$r(t,s) = E\{[X(t) - m(t)][X(s) - m(s)]\} = \sigma^2 e^{-\alpha(s-t)},\ t \leq s.$$

The transition density function $(t < s)$ is

$$\begin{aligned} p(x,t;y,s) &= \frac{\partial}{\partial y} P\{X(s) \leq y \mid X(t) = x\} \\ &= [2\pi A(t,s)]^{-1/2} \exp\left[-\left(\frac{\{y - x\exp[-\alpha(s-t)]\}^2}{2A(t,s)}\right)\right], \end{aligned}$$

where

$$A(t,s) = \sigma^2\{1 - \exp[-2\alpha(s-t)]\},$$

$\sigma^2 > 0$ and $\alpha > 0$. This yields conditional mean and variance functions

$$E\{X(s) \mid X(t) = x\} = \int_{-\infty}^{\infty} yp(x,t;y,s)dy = xe^{-\alpha(s-t)},$$

and

$$\mathrm{Var}\{X(s) \mid X(t) = x\} = A(t, s).$$

The correlation between $X(t)$ observations separated by one unit of time is $e^{-\alpha}$. Thus, $e^{-\alpha}$ can be regarded as the theoretical autocorrelation function with lag 1. An observed series can be used to calculate the sample autocorrelation coefficient with lag 1:

$$r_1 = \frac{\sum_{1=2}^{n}(X_i - \overline{X})(X_{i-1} - \overline{X})}{\sum_{i=1}^{n}(X_i - \overline{X})^2},$$

where $\overline{X} = \frac{1}{n}\sum_{i=1}^{n} X_i$.

3. APPLICATION TO SPOT RATES

Let us now apply an O.U. process to the spot rate description. We denote the long term mean of the $r(t)$ process by γ, and assume that time 0 is chosen so that $r(0) = \gamma$. Let $X(t) = r(t) - \gamma$, $0 \leq t < \infty$. Then $X(0) = 0$, and $X(t)$ represents random deviations from γ, as time flows beyond our chosen initial time 0. The conditional mean function is

$$E\{X(s) \mid X(t) = x - \gamma\} = (x - \gamma)e^{-\alpha(s-t)}.$$

But

$$E\{X(s) \mid X(t) = x - \gamma\} = E\{r(s) - \gamma \mid r(t) = x\} = E\{r(s) \mid r(t) = x\} - \gamma.$$

Thus

$$E\{r(s) \mid r(t) = x\} = \gamma + (x - \gamma)e^{-\alpha(s-t)},$$

for $t \leq s$, in agreement with equation (25) of Vasicek (1977). The conditional variance function is

$$\mathrm{Var}\{X(s) \mid X(t) = x - \gamma\} = A(t, s).$$

But

$$\mathrm{Var}\{X(s) \mid X(t) = x - \gamma\} = \mathrm{Var}\{r(s) - \gamma \mid r(t) = x\} = \mathrm{Var}\{r(s) \mid r(t) = x\}.$$

Thus

$$\mathrm{Var}\{r(s) \mid r(t) = x\} = \sigma^2\{1 - \exp[-2\alpha(s - t)]\}$$

for $t \leq s$, in agreement with equation (26) of Vasicek (1977), if we let $\sigma^2 = \rho^2/(2\alpha)$.

4. SELECTED PROBABILITIES FOR O.U. PROCESSES

It would be valuable to know the probabilities that the deviations $X(t)$ from γ would exceed preassigned bounds during various time intervals. Knowledge of such probabilities is needed for some applications of immunization theory. Consider probabilities of the $X(t)$ process exceeding various multiples of σ during T time units,

$$P\{\max_{0\le t\le T} X(t) > A\sigma \mid X(0) = 0\}.$$

These probabilities, or actually their complements

$$P\{\max_{0\le t\le T} X(t) < A\sigma \mid X(0) = 0\}$$

were calculated by Beekman and Fuelling (1977) by approximating an integral equation through a set of equations. The technique was patterned after one used by Park and Schuurmann (1976), and used a transformation of probabilities concerning the tied down O.U. process into ones connected with the Wiener process. For the probabilities discussed by Beekman and Fuelling (1977), it was assumed without any loss of generality that $\alpha = 1$ and $\sigma = 1$. If $\{Y(t),\ 0 \le t < \infty\}$ is an O.U. process with $\alpha = 1$ and $\sigma = 1$, then

$$P\{\max_{0\le t\le T} X(t) > A\sigma \mid X(0) = 0\} = P\{\max_{0\le t\le \alpha T} Y(t) > A \mid Y(0) = 0\}.$$

The following table is obtainable from Beekman and Fuelling (1977).

αT	A:	0.5	1.0	2.0	3.0
6		0.991340	0.925174	0.434262	0.059290
7		0.995466	0.949169	0.486512	0.070197
8		0.997626	0.965468	0.533930	0.080975
9		0.998757	0.976541	0.576968	0.091627
10		0.999349	0.984064	0.616031	0.102156

Another source for tables of passage time distributions for O.U. processes is Keilson and Ross (1976).

Let us now assume that explicit provision for growing deviations is included in the actuarial or investment assumptions and that the constant boundary $A\sigma$ is replaced by $A\sigma(1+R)^t$, $0 \leq t \leq T$. The following table from Beekman and Fuelling (1979) gives the probabilities of exceeding such a boundary during T time units, when $R = 0.04$.

If one compares values in the two tables for $\alpha T = 7$ and $\alpha T = 10$, it reveals the diminution in probabilities of adverse deviations when explicit provision is made for growing deviations, perhaps because of inflation.

$$P\{\max_{0\leq t\leq T}[X(t) - A\sigma(1+0.04)^t] > 0 \mid X(0) = 0\}$$

αT	A: 0.5	1.0	2.0	3.0
1	0.759225	0.443469	0.062135	0.002353
5	0.979892	0.856340	0.265997	0.018135
7	0.993472	0.917020	0.313317	0.020125
10	0.998654	0.958556	0.350675	0.020898

Beekman and Fuelling (1979) have also given such tables for $R = 0.03, 0.05, 0.06$, and 0.07. Such tables permit various comparisons. Let us fix $A = 3$, but vary αT and R.

$$P\{\max_{0\leq t\leq T}[X(t) - 3\sigma(1+R)^t] > 0 \mid X(0) = 0\}$$

R	αT: 1	5	7	10
0.03	0.002571	0.023147	0.027169	0.029626
0.04	0.002353	0.018135	0.020125	0.020898
0.05	0.002153	0.014277	0.015191	0.015399
0.06	0.001970	0.011321	0.011710	0.011758
0.07	0.001801	0.009059	0.009213	0.009222

For fixed αT, the probabilities of adverse deviations decrease as R increases. This is consistent, of course, as greater conservatism is present for bigger R values. For fixed R, the probabilities increase as αT grows. This reflects the greater chance of a sample function exceeding a boundary as the observation period grows.

These probabilities were calculated by approximating an integral equation through a set of equations. Again, the technique was patterned after one used by Park and Schuurmann (1976), and used a transformation of probabilities concerning the tied down O.U. process into ones connected with the Wiener process. Theorem 1 of Beekman and Fuelling (1979) again relates the general O.U. process with one for which $\sigma^2 = 1$ and $\beta = 1$. Its conclusion is the equation above our first table.

ACKNOWLEDGMENT

I would like to acknowledge the help of Ball State University in the preparation of this paper through an Academic Year Research Grant.

REFERENCES

Beekman, J. A. (1974), "A new collective risk model." *Transactions of the Society of Actuaries* **25**, 573–589.

Beekman, J. A. (1975, 1976), "Compound Poisson processes as modified by Ornstein-Uhlenbeck processes, Parts I and II." *Scandinavian Actuarial Journal*, 226–232, 30–36.

Beekman, J. A., and C. P. Fuelling (1977), "Refined distributions for a multi-risk stochastic process." *Scandinavian Actuarial Journal*, 177–183.

Beekman, J. A., and C. P. Fuelling (1979), "A multi-risk stochastic process." *Transactions of the Society of Actuaries* **30**, 371–397.

Beekman, J. A., and C. P. Fuelling (1980), "Simulation of a multi-risk collective model." In *Computational Probability*, ed. P. M. Kahn, pp. 287–301. New York: Academic Press.

Boyle, P. P. (1978), "Immunization under stochastic models of the term structure." *Journal of the Institute of Actuaries* **105**, 177–187.

Boyle, P. P. (1980), "Recent models of the term structure of interest rates with actuarial applications." *Transactions, 21st International Congress of Actuaries* **4**, 95–104.

Brennan, M. J., and E. S. Schwartz (1977), "Savings bonds, retractable bonds and callable bonds." *Journal of Financial Economics* **5**, 67–88.

Cox, J. C., J. E. Ingersoll, Jr., and S. A. Ross (1981), "A reexamination of traditional hypotheses about the term structure of interest rates." *Journal of Finance* **36**, 769–799.

Keilson, J., and H. Ross (1976), "Passage time distributions for the Ornstein-Uhlenbeck process." *Selected Tables in Mathematical Statistics* **3**, American Mathematical Society, Providence, Rhode Island.

Park, C., and F. J. Schuurmann (1976), "Evaluations of barrier crossing probabilities of Wiener paths." *Journal of Applied Probability* **13**, 267–275.
Richard, S. F. (1978), "An arbitrage model of the term structure of interest rates." *Journal of Financial Economics* **6**, 33–57.
Vasicek, O. (1977), "An equilibrium characterization of the term structure." *Journal of Financial Economics* **5**, 177–188.

S. Broverman [1]

A NOTE ON VARIABLE INTEREST RATE LOANS

1. INTRODUCTION

Interest rates as random variables have been considered in a number of contexts. In particular, Boyle (1976) and Panjer and Bellhouse (1980) have analysed present and accumulated values of annuities-certain and life annuities when the valuation rate of interest is a random variable. In this paper the complete amortization of an annuity-certain (or variable rate loan—VRL) is analyzed when the valuation rate is a random variable.

One of the advantages of a fixed rate loan is the certainty over the term of the loan of the size of the payment. With a VRL, the payment size will vary to some extent with the fluctuation of the interest rate. This paper presents a method of determining the payment for a VRL that results in less expected variability in the series of loan payments than would occur in the typically used method of calculating the loan payment (under appropriate circumstances the method presented minimizes the expected variability in the sequence of payments). The effects of this method on other aspects of the amortization such as outstanding balances and interest paid are also investigated.

A fixed rate n-payment loan of amount 1 at periodic effective interest rate i (or force of interest δ) has level periodic payment $R = 1/a_{\overline{n|}i} = \frac{i}{1-(1+i)^{-n}}$. The payment required on a VRL will depend upon

(a) the joint distribution of the interest rates $(i_1, i_2, \ldots, i_n)$ (or forces of interest $(\delta_1, \ldots, \delta_n)$ for the n-period term), and

(b) the method of payment calculation.

The typical payment calculation is to set the VRL payment to be $R_0 = 1/a_{\overline{n|}\overline{E}}$, where $\overline{E} = (\overline{i}_1, \ldots, \overline{i}_n)$ is the vector of expected rates for the n periods (there are various other ways of defining the VRL payment that do not give exactly the same R_0, and in practice a premium over the fixed interest rate for the same period is often incorporated into $\overline{i}$). This is the

[1] Department of Statistics, University of Toronto, Toronto, Ontario M5S 1A1

I. B. MacNeill and G. J. Umphrey (eds.), Actuarial Science, 165–172.

method of VRL payment calculation that will be used for comparing with the method introduced.

2. INDEPENDENT RATES OF INTEREST

Suppose that a payment scheme with tth payment R_t has been chosen for an n-payment VRL. First consider the case of an n-payment VRL for which the final (nth) payment is exactly enough to retire the loan. The amount of the final payment will be a random variable R_n given by the expression

$$R_n = \prod_{t=1}^{n} e^{\delta_t} - \sum_{t=1}^{n-1} R_t \cdot \left(\prod_{s=t+1}^{n} e^{\delta_s} \right).$$

The measure of smoothness for the series of payments will be

$$S^2 = \sum_{t=1}^{n} (R_t - \overline{R})^2.$$

The objective is to minimize $E(S^2)$.

In this section of the paper it is assumed that the periodic forces (and effective rates) of interest are independent and identically distributed (with mean $\bar{i}$). Begin with the assumption that R_t has a fixed value R for $t = 1, 2, \ldots, n-1$, with R_n being the only variable payment. Under these assumptions it follows that

$$E[S^2] = \left(\frac{n-1}{n} \right) \cdot (R^2 - 2R \cdot E[R_n] + E[R_n^2]).$$

Define $A = E[e^{\delta_t}] = E[1 + i_t]$ and $B = E[e^{2\delta_t}] = E[(1 + i_t)^2]$. Then

$$E[R_n] = A^n - R \cdot \left(\frac{A^n - A}{A - 1} \right),$$

and

$$E[R_n^2] = B^n - 2 \cdot \left[\frac{AB^n - A^n B}{B - A} \right] + R^2 \cdot \left[\left(\frac{B^n - B}{B - 1} \right) \cdot \left(\frac{B + A}{B - A} \right) - \left(\frac{A^n - A}{A - 1} \right) \cdot \left(\frac{2B}{B - A} \right) \right]$$

and $E[S^2]$ becomes $\left(\frac{n-1}{n}\right) \cdot (aR^2 + bR + c)$, where

$$a = 1 + \left(\frac{B^n - B}{B - 1}\right) \cdot \left(\frac{B + A}{B - A}\right) - \left(\frac{A^n - A}{A - 1}\right) \cdot \left(\frac{2A}{B - A}\right),$$

$$b = -2A \cdot \left[\frac{B^n - A^n}{B - A}\right],$$

$$c = B^n.$$

Thus, in this case, $E[S^2]$ is a quadratic function of R. The minimum occurs when $R = -b/(2a)$. It can be shown that if $B = A^2$ (i.e., if δ_t - and i_t - are not random) then $b^2 - 4ac = 0$ and $R = 1/a_{\overline{n}|\bar{i}}$ (the usual fixed rate amortization payment at effective rate $\bar{i}$ per period) and $S^2 = 0$ ($R_n = R$). If $B > A^2$ then $b^2 - 4ac < 0$, the minimum of $E[S^2]$ occurs at $R = -b/2a > 1/a_{\overline{n}|\bar{i}}$, and $E[S^2]$ with $R' = -b/(2a)$ is less than $E[S^2]$ with $R = 1/a_{\overline{n}|\bar{i}}$ (the periodic payment that would be calculated on the basis of the standard method outlined in Section 1).

Table 1 presents a numerical example with $n = 10$ and δ having a normal distribution with mean .1 and standard deviation .02887. Method 1 in the table refers to $R = 1/a_{\overline{10}|.1}$ for the first nine payments, and Method 2 refers to $R' = -b/(2a)$ for the first nine payments. The initial amount of the loan is 1,000. The table shows the slight reduction in $E[S^2]$ when Method 2 is used, and the difference in payment amount between Methods 1 and 2.

A simulation of 100,000 ten-period sequences was run and the rest of Table 1 outlines the results. As expected, the sample mean for the tenth payment is smaller for Method 2 than for Method 1 (since the first nine payments in Method 2 are larger than in Method 1). The second column gives the sample variance.

The simulated estimates for $E[S^2]$ are also indicated in the table along with the square of the standard error of the estimate. The rest of the table compares outstanding balances, interest per period and total interest paid (along with sample variances) generated by the simulation. It appears that Method 2 results in smaller variances in all these quantities.

In practice, there will be some sort of periodic updating of the payment based on the rates of interest actually experienced. Considering the case in which at each payment point t the payment is recalculated on the basis of an $(n - t + 1)$-payment loan for amount OB_{t-1} (the previous period's outstanding balance) at the most recent rate (or force) of interest i_t, the sequence of payments has the form

$$R_t = \frac{OB_{t-1}}{a_{\overline{n-t+1}|i_t}}.$$

Table 1. *Standard amortization payment per* 1000 *is* 166.703172983, *and the expected sum of squared deviations is* 17383.6527831. *Payment for minimum sum of squared deviations is* 166.92492287 *and the expected sum of squared deviations is* 17371.7504954.

Method 1		Method 2	
Annual payment for first 9 years			
166.703173		166.924923	
Tenth year payment and sample variance			
165.7841	19388.1770	162.3740	119361.6953
Variation in payments and in estimate			
17449.9616	6725.3278	17444.0079	6624.3769
Outstanding balances and sample variances			
938.5166	1027.0280	938.2949	1027.0280
870.8521	2158.5185	870.3852	2158.0920
796.1054	3399.9864	795.3674	3398.6460
713.6202	4813.5852	712.5824	4810.7381
622.3167	6421.2069	620.9475	6416.1875
521.3675	8260.2354	519.6319	8252.3473
409.7704	10382.9912	407.6296	10371.5012
286.3657	12862.4771	283.7769	12846.6888
149.9438	15830.6701	146.8597	15809.7395
.0000	.0000	.0000	.0000
Yearly interest and variance estimates			
105.2198	1027.0280	105.2198	1027.0280
99.0386	916.4648	99.0152	916.0378
91.9564	794.7740	91.9071	793.9445
84.2180	693.3715	84.1399	692.1472
75.3996	580.4863	75.2900	578.9368
65.7540	474.1445	65.6093	472.3455
55.1061	379.9253	54.9226	377.9752
43.2985	296.6417	43.0722	294.7236
30.2813	241.3946	30.0077	239.6858
15.8403	215.1045	15.5143	213.9287
Total annual interest and variance estimates			
666.1127	19387.9831	664.6983	19361.5017

Table 2 outlines the result of using this updating method (called Method 3) with the same simulation used to produce Table 1. There is a large reduction in variation in all quantities (particularly S^2) compared with Methods 1 and 2. The minimization technique of Method 2 can now be adapted to the updating of Method 3 to produce Method 4 in which at each time t, R_t is calculated on the basis of Method 2 with A and B the same, the previous outstanding balance (OB_{t-1}), and with term $n-t+1$. The results are shown, again with the same simulation, in Table 3. Note that the estimate of S^2 is about one-half that for Method 3. However, variation in payment amount is greater in the later payments for Method 4 than for Method 3, and variation in other quantities is also greater than in Method 3.

It appears that Method 4 provides a reasonable way with which to take advantage of a VRL while reducing the risk of fluctuating payments.

3. GENERAL INTEREST RATE DISTRIBUTIONS

Expressions (1) and (2) of Section 2 also apply to the more general setting in which periodic rates are not necessarily independent (they are often modelled as a time series). This also results in the representation of $E[S^2]$ in the quadratic form aR^2+bR+c, where a, b and c are now combinations of moment generating functions. In the analysis done by Panjer and Bellhouse (1980), the expression $\Delta(t)$ is defined to be the sum $\delta_1+\delta_2+\cdots+\delta_t$. If $\nabla(t)$ is defined to be the sum $\delta_{t+1}+\delta_{t+2}+\cdots+\delta_n$ then the coefficients of the quadratic for $E[S^2]$ can be expressed as

$$a=1+2\cdot\sum_{t=1}^{n-1}M_{\nabla(t)}(1)+\sum_{t=1}^{n-1}M_{\nabla(t)}(2)+2\cdot\sum_{t=1}^{n-2}\sum_{u=t+1}^{n-1}M_{\nabla(u)+\nabla(t)}(1),$$

$$b=-2\cdot\left[M_{\nabla(0)}(1)+\sum_{t=1}^{n-1}M_{\nabla(0)+\nabla(t)}(1)\right],$$

and

$$c=M_{\nabla(0)}(2).$$

Based on the methods outlined by Panjer and Bellhouse (1980), a, b and c can be found for various time series that are used for modelling interest rates. Then Method 4 of Section 2 can be applied by updating at each payment point, based on the observed interest rates and the assumed time series distributions of future interest rates.

Table 2. *Method 3*

Annual payments and variances	
167.070263	512.735435
167.061053	441.718185
166.897919	369.999346
166.817667	311.988732
166.573961	251.522075
166.416929	197.897635
166.292863	151.515051
166.169732	111.083527
166.078866	80.700504
166.012203	58.273779
Variation in payments and variance of estimate	
1869.931213	12.147395
Outstanding balances and variances	
938.149544	89.643269
870.089078	160.503498
795.070932	208.489993
712.360842	233.553310
621.048256	232.250053
520.248113	203.995213
408.944381	153.193142
285.993903	89.069040
150.143507	28.941492
.000000	.000000
Annual interest and variances	
105.219807	1027.028015
99.000586	904.094102
91.879773	769.998441
84.107577	652.977465
75.261375	521.422966
65.616786	396.587506
54.989131	279.967689
43.219254	172.089240
30.228470	84.642550
15.868696	23.147747
Total interest paid and variance	
665.391456	6189.645005

Table 3. *Method 4*

Annual payments and variances	
166.924923	.000000
166.790892	32.452507
166.711880	65.546260
166.645466	98.965900
166.165043	134.555877
166.560959	172.480899
166.506925	213.092164
166.460498	258.959733
166.409458	316.260183
166.371968	445.639437
Variation in payments and variance of estimate	
930.076872	4.705882
Outstanding balances and variances	
938.294884	1027.028015
870.519235	1787.193443
795.729161	2256.467714
713.260732	2465.874000
622.006832	2405.383157
521.164265	2087.623294
409.741500	1569.024646
286.582267	937.966077
150.468240	345.327987
.000000	.000000
Annual interest paid and variances	
105.219807	1027.028015
99.015423	916.037766
91.921806	789.784492
84.177036	678.708433
75.360693	550.966425
65.718841	424.223705
55.084161	303.998627
43.301265	189.786365
30.295431	95.622308
15.903729	27.132780
Total interest paid and variance	
665.998013	10033.017132

REFERENCES

Boyle, P. P. (1976), "Rates of return as random variables". *Journal of Risk and Insurance* **43**, 693–713.

Panjer, H. H., and D. R. Bellhouse (1980), "Stochastic modelling of interest rates with applications to life contingencies". *Journal of Risk and Insurance* **47**, 91–110.

Patrick L. Brockett [1] and Naim Sipra [2]

LINEARITY AND GAUSSIANITY OF INTEREST RATE DATA: AN EMPIRICAL TIME SERIES TEST

1. INTRODUCTION

It is generally accepted that people's economic-choice behavior is motivated by the utility they derive from actual or expected consumption of real goods. The ratio of marginal utility from any such consumption today as opposed to the marginal utility of future consumption leads to real interest rates, and to relative prices denominated in units of consumption goods (or, utils per consumption). But, trading is done with prices given in terms of money. This intertemporal preference for consumption denominated in monetary terms is the basis for interest rate behavior. Thus, while consumption unit prices reflect people's choices regarding intertemporal consumption behavior, in order to study the *true* consumption choices one needs to know how the prices denominated in consumption units are related to the money prices.

The stochastic behavior of interest rates, which reflects variations in the above mentioned relationship, is of considerable importance in finance, economics, and actuarial science. In particular, the relationship between the real rate of interest, the nominal rate of interest quoted in the market place, and the rate of inflation is of interest to those making financial investment decisions such as actuaries and pension fund managers. There have been several attempts to model interest rate series in the actuarial and financial literature. Of particular note are the papers of Cox *et al.* (1977), Boyle (1978), Panjer and Bellhouse (1980), and Bellhouse and Panjer (1981). In these papers the series are often modeled by an autoregressive, autoregressive moving average, or a continuous time diffusion model. In all these cases the

[1] Department of Finance and Applied Research Labs, University of Texas at Austin, Austin, Texas 78712

[2] Department of Finance, University of Colorado at Denver, Denver, Colorado 80202

I. B. MacNeill and G. J. Umphrey (eds.), Actuarial Science, 173–183.

series can be transformed into a so-called linear time series model which involves, perhaps, certain regressive covariates. In this paper we use a newly developed time series test for linearity and/or Gaussianity of time series which is based upon the bispectrum of the series to test the above mentioned linear time series relationship. We conclude that in most cases these series are, in fact, nonlinear and non-Gaussian.

The remaining paper is organized as follows: In Section 2 we develop, briefly, the theory of linear and Gaussian time series. In Section 3 we discuss the logic behind the Hinich time series tests for linearity and normality of a time series which we use subsequently to test various series derived from interest rate data. The results of these hypothesis tests are reported in Section 4.

2. LINEAR AND NONLINEAR TIME SERIES

Many of the models used in finance and economics assume a linear generating process of the type below:

$$Y = \alpha + \sum \beta_i X_i + \epsilon. \tag{1}$$

If the generating process is stationary over time one can write (1) as a time series, or

$$Y(t) = \alpha + \sum \beta_i X_i(t) + \epsilon(t). \tag{2}$$

If the process in (2) is properly specified for the time series statistical tests which most authors perform, then $\epsilon(t)$ follows a purely random mean zero process. This is because after accounting for the proper covariates and model, there should be no structural dependencies left in the series. We define a stationary time series as purely random if $\epsilon(t_1), \epsilon(t_2), \ldots, \epsilon(t_n)$ are independent random variables with zero mean. Since, typically, the error terms are assumed to follow a normal process, it is often considered sufficient to test the error terms for zero autocorrelation to conclude that $\epsilon(t)$ are independent. If, however, $\epsilon(t)$ are not normal, then even if $E[\epsilon(t)\epsilon(t-k)] = 0$ for all $k = 0$ it is possible that $\epsilon(t)$ are not independent, in which case (2) with purely random $\epsilon(t)$ cannot be said to depict the true generating process. There may be additional exploitable information to be found in residuals. In order for $\epsilon(t)$ to be independent it is not sufficient that the second-order cumulant $E[\epsilon(t)\epsilon(t-k)]$ is zero, but rather all higher cumulants must also be zero. Hinich (1982) derived a test for the linearity and Gaussianity of a time series using these higher order cumulants, which we present subsequently.

A time series is said to be generated by a linear process if it can be expressed as follows:

$$X(n) = \sum a(m)u(n-m), \tag{3}$$

where $\{u(n)\}$ is purely random. When $a(m) = 0$ for $m < 0$ and $m > q$, then

$$X(n) = \sum_{m=0}^{q} a(m)u(m-n) \tag{4}$$

is called a one-sided (or causal) moving average of order q. The independence of $u(n)$ variates makes a linear process special. If the process is linear and a least squares regression is applied to $X(n)$ with (or without) the proper number of independent explanatory variables, the error term will be a linear process. If, however, the process is nonlinear, but is modeled as a linear process, the intrinsic nonlinearity will transfer to the error term. Thus, by testing the residuals from such a regression for nonlinearity, one can determine whether a time series was generated by a stationary linear or nonlinear innovation process. Note that if the linearity of the residuals is rejected then one must also reject the normality assumption for the residuals. However, a rejection of the normality is not automatically a rejection of linearity.

The past fifteen years have witnessed a tremendous growth in the application of linear time series analysis to real data. Applied time series analysis is now a standard feature of undergraduate and graduate curriculums in engineering, the physical sciences, life sciences, social sciences, and business education. This phenomenon is a product of three related events: (1) the prodigious research effort directed at linear time series analysis, (2) a realization that many scientific problems are amenable to time series analysis, and (3) rapid growth in the number of time-based machine-readable data sets, and the availability of linear time series methods on standard statistical software packages. It must be pointed out, with emphasis, that there is no particular reason why empirical time series should conform to linear time series models, or even be well approximated by a linear model. The purpose of this paper is to present statistical techniques helping to determine which time series are actually linear processes, and which time series are not amenable to linear series modeling. We then apply these time series tests to some interest rate time series which have previously been modeled using linear time series methods.

The stationary autoregressive (AR), and the stationary autoregressive-moving average (ARMA) models which are widely used in time series analysis, are finite order representations of the general causal linear model (Priestley, Sec. 7.7, 1981). Even if the data appears to be nonstationary during the observation period, linear filtering techniques such as differencing (Box and Jenkins, Sec. 4.1, 1966) and trend regression can often be used to transform the data into a seemingly stationary sequence.

If the observed process is the output of a nonliner operation on an input process, then the sample autocovariances are insufficient for identifying the structure of the nonlinear filter. For example, suppose that

$$x(t) = u(t) + au(t-1)u(t-2),$$

where $\{u(t)\}$ is a *purely random* process; that is, the $u(t)$ are independent zero mean variates. Then it is easy to check that the covariance

$$E[x(t)x(t+m)] = 0 \text{ for all } m \neq 0.$$

Thus $\{x(t)\}$ is a stationary white noise process with *nonlinear* structure that yields dependencies over three time points.

Before describing the Hinich test for linearity and normality used in this paper, we first discuss the importance of such tests.

The fact that certain series studied in finance are not Gaussian has been known for a while. However, the majority of the tests have been performed on the marginal distributions, whereas, the common assumption for normality is for the joint distribution. Since it is possible for the marginal distributions to be normal while the joint distribution is not normal, then one may accept normality as an approximation based on tests on the marginal distribution but not accept this hypothesis if the tests had been performed on the joint distribution. As will be shown later, the assumption of normality (or even linearity) for interest rate series is a bad approximation based on joint normality. If the residuals are nonnormal then most of the hypothesis tests reported related to these series are suspect. Moreover, in such a situation, the standard linear econometric techniques must be augmented to account for nonlinear stochastic variation before meaningful modeling may be accomplished. Subba Rao and Gabr (1984) gave a computer code for fitting a bilinear (nonlinear) model to the residual errors after having fitted an ARMA model to the data.

Next we discuss how an incorrect assumption regarding normality can affect the tests we use to support economic hypotheses. An oft quoted result of market efficiency is that prices (and other well chosen time series) follow a random walk. A time series is a random walk if $\{X(n) - X(n-1)\}$ is a purely random series. Typically, in the finance literature it has been deemed sufficient to test the auto-correlation of the time series and if it is zero for all positive lags then the time series is assumed to follow a random walk. The above conclusion is correct provided that the time series follows a normal distribution. For a further discussion of this problem in the context of stock price series, see Hinich and Patterson (1984).

The linearity of time series involving interest rates (e.g., autoregression, ARMA, etc.) is rejected using the tests presented subsequently. This is the first time interest rate time series have been tested for linearity.

3. THE HINICH TEST FOR LINEARITY AND NORMALITY

The essential calculations used in the Hinich test for linearity and normality of a time series involve examining the third order cumulants of the time series. For statistical work, estimates of third order cumulants are very cumbersome to deal with, but the double Fourier transform of the third order cumulant, called the bispectrum, is much easier to handle. Subba Rao and Gabr (1980), and Hinich (1982), present statistical tests for determining whether a given stationary time series $\{X(n)\}$ is linear. It is possible that $\{X(n)\}$ is linear without being Gaussian, but all the stationary Gaussian time series are linear. Both of the above papers also give time series tests for joint Gaussianity, and the tests presented in both papers are based upon the sample bispectrum of the time series. The Hinich test is non-parametric, and is robust. Accordingly, the tests presented in this paper use the Hinich test.

Let $\{X(n)\}$ be a stationary time series and assume without loss of generality that $E[X(n)] = 0$. The spectrum of $\{X(n)\}$ is the Fourier transform of the autocovariance function $C_{XX}(n) = E[X(t+n)X(t)]$:

$$S(f) = \sum_{n=0}^{\infty} C_{XX}(n) \exp\{-2\pi i f n\}.$$

Many papers use the spectrum $S(f)$ as a way to examine the correlation structure of $X(n)$. See Granger and Morgenstern (1963) for numerous applications of spectral analysis techniques to finance. In particular, $X(n)$ is serially uncorrelated (white noise) if $S(f)$ is constant.

The *bispectrum* of $\{X(n)\}$ is defined to be the (two dimensional) Fourier transform of the third moment function

$$C_{XXX}(n:m) = E[X(t+n)X(t+m)X(t)],$$

$$B(f_1, f_2) = \sum_{m}\sum_{n} C_{XXX}(n,m) \exp\{-2\pi i f_1 n - 2\pi i f_2 m\}.$$

The bispectrum is a spatially periodic function whose principal domain is the triangular set

$$\Omega = (0 < f_1 < 1/2; \;\; f_2 < f_1; \;\; 2f_1 + f_2 < 1).$$

A rigorous treatment of the bispectrum is given by Brillinger and Rosenblatt (1967).

The third order cumulants are expressed in terms of the bispectrum by the inverse fourier transform

$$C_{XXX}(m,n) = \int\int_{\Omega} B(f_1, f_2) \exp[i2\pi(f_1 m + f_2 n)] df_1 df_2. \tag{6}$$

Suppose $\{X(t)\}$ is a linear time series, that is, it can be expressed as in (3). Assuming that $\{X(t)\}$ is a third order stationary time series, that is, the joint distribution of $X(n_1), \cdots, X(n_N)$ depends only on the time differences between the n_k's, and assuming also that all moments exist for the series, third-order stationarity implies that the mean $\mu_X = E[X(n)]$ and that the covariance C_{XX} and the third order cumulant function $C_{XXX}(r,s) = E[X(n+r)X(n+s)X(n)]$ are independent of n. For simplification we will assume throughout that $\mu_X = 0$ (this can be accomplished by centering the series). Assuming also that $X(t)$ is summable, that is that $\sum a(n)$ is finite, the bispectrum of a linear series $\{X(t)\}$ can be shown to be (Brillinger, 1975)

$$B(f_1, f_2) = \mu_3 A(f_1) A(f_2) A^*(f_1 + f_2) \tag{7}$$

where

$$\mu_3 = E[u^3(t)],$$

$$A(f) = \sum a(n) \exp(-ifn), \tag{8}$$

and $A^*(f)$ is its complex conjugate. Since in the linear process case the spectrum of $X(t)$ can be shown to be

$$S_x(f) = \sigma_u^2 \mid A(f) \mid^2,$$

it follows from (6) and (7) that

$$\frac{\mid B(f_1, f_2) \mid^2}{S_x(f_1) S_x(f_2) S_x(f_1 + f_2)} = \frac{\mu_3^2}{\sigma_u^6} \tag{9}$$

for all f_1, f_2 in Ω.

With the stationary assumptions, the right hand side of (9) is a constant. The left hand side of (9) is the square of the skewness function of $\{u(t)\}$. If $X(t)$ was composed of normal variables then $\mu_3 = 0$. Using the fact that the skewness function for a linear time series is zero for a Gaussian series and a constant otherwise, Hinich derived a statistical test for linearity and the Gaussianity of a time series. He obtains an estimate of the bispectrum, $\hat{B}(f_1, f_2)$, and of the spectrum, $\hat{S}(f)$, and then estimates the ratio in (9) at different frequency pairs (f_1, f_2) in the principal domain by $|\hat{B}(f_1,f_2)|^2 / [\hat{S}(f_1)\hat{S}(f_2)\hat{S}(f_1+f_2)]$. If these ratios differ too greatly over different frequency pairs, he rejects the constancy of the ratio, and hence the linearity of the time series $\{X(n)\}$. If the estimates differ too greatly from zero, he rejects the Gaussianity time series model.

The test statistic he derives for testing linearity is based upon the interquartile range of the estimated ratio over the set of pertinent frequency pairs. If the ratio in (9) is constant, then the interquartile range is small. If it is not constant, then the interquartile range is larger. A discussion of the calculations involved and precise formulae are given by Hinich (1982). A more intuitive explanation of the tests is given by Hinich and Patterson (1985) and will not be repeated here.

A comment about the stationarity of the time series is in order. Since the test statistics were derived under the assumption of third order stationarity for the time series, one might question the validity of the tests performed in this paper involving time series over a thirty-year period. Somewhat heuristically, the tests for linearity and Gaussianity can be described as searching for statistically significant linkages between various observations of the time series. For example, the time series

$$X(n) = 0.7X(n-3) + 0.5X(n-7)X(n-8)$$

has nonlinear linkages between observations. If the time series exhibit certain types of nonstationarity it will become difficult to detect such linkages and consequently that much more difficult to reject linearity and normality (e.g., the bispectrum will be smeared somewhat and flattened because of the logs of phase coherence caused by the nonstationarity). Thus, in most economic time series, nonstationarity should tend to make the test somewhat conservative. Moreover, nonlinear series will exhibit time path behavior which, when forced into a linear time series model for stationarity testing, may appear erroneously to be nonstationary. All current tests for stationarity of time series are based on linear structure and will fail to be valid in certain nonlinear models. Thus, the nonlinear structure of the model should be viewed as a potential alternative to linear models, but one which cannot be tested using only the standard statistical methods from linear time series analysis.

Finally, what does it mean if we do find that residuals of a linear regression form a nonlinear time series? It could mean that there exists some systematic nonlinear relationship in the dependent variable part that is not being explained by the independent variables. The troublesome part, of course, is the source of this nonlinearity. There are two candidates for this source. One, that there really is a nonlinear relationship that explains part of the variability of the dependent variable, but which is being ignored. The other is that we are measuring the variables with errors, and the errors are nonlinear series. Additionally, the standard methods for estimating the coefficients of a *linear* model will be biased and even the best fitting of the potential linear models not necessarily will be selected. To see the significance of the above problem to hypothesis testing of interest rates, consider the following example.

We wish to test the hypothesis that the real interest rate is equal to the nominal interest rate minus the inflation rate. To test the above hypothesis we run the regression:

$$r_t = \alpha + \beta_1 R_t + \beta_2 \Delta_t + \epsilon_t,$$

where

r_t = monthly growth rate in real per capita consumption,
R_t = nominal holding-period return on 1-month T-bills, and
Δ_t = monthly inflation rate.

If $\alpha = 0$, $\beta_1 = 1.0$ and $\beta_2 = -1.0$ then we would claim that we cannot reject the hypothesis that real rate = nominal rate - inflation rate. Accordingly, if $\alpha \neq 0$, or $\beta_1 \neq 1.0$ or $\beta_2 \neq -1.0$ (at say the 95% confidence level) then we would reject the above null hypothesis. The above technique of hypothesis testing is quite standard in economics and finance. However, an important assumption for the validity of the above argument is being ignored. For the above mentioned hypothesis test to be valid the error term should be a zero-mean purely random process.

Thus, if we test residuals from the above regression and find them to be nonlinear then either our assumption about the residuals being zero-mean purely random process was incorrect (possibly due to measurement errors or proxy problems)—in which case regardless of the values for the estimates of α, β_1 and β_2 we can neither reject nor not-reject the null hypothesis using standard statistical tests—or our assumption about the error term was correct, but the true relationship between the real rate, the nominal rate and the inflation rate was not as posed in the null hypothesis. We cannot tell which of these two cases is true from the test, and hence, if the model is incorrect, the inference drawn from the model is suspect.

4. TESTABLE HYPOTHESES INVOLVING INTEREST RATES

In addition to the obvious hypotheses that nominal, real, and inflation rate time series follow linear time series models there are two other tests we shall perform.

A. Constant real return

The expected real rate of return is a constant. That is,

$$r_t = E(r_t) + \epsilon_t, \tag{10}$$

where r_t is the real rate of return and ϵ_t is purely random.

The above is equivalent to saying that the real rate follows a random walk. The assumption of constant real rate is difficult to defend, especially if the test is over a twenty-year period, however there are some advocates of this position, and as Fama (1975) says, let the data decide!

B. Nominal interest rates as predictors of inflation

If the expected real rate is a constant then all the variation through time in the nominal rate of return, R_t, is a direct manifestation of the variation in the expected inflation rate, $E(\Delta_t)$, as assessed by the market. That is, R_t is the best predictor for $E(\Delta_t)$. The hypothesis that $E(r_t)$ is a constant and that nominal rates reflect all information about the expected inflation rate can be tested by running the following least-squares regression:

$$\Delta_t = \alpha + \beta R_t + \epsilon_t. \tag{11}$$

If ϵ_t follows a purely random process and $\beta = +1.0$ then one cannot reject the hypothesis that $E(r_t)$ is a constant and that R_t is the best predictor of deflation.

The data used in this test involved the yields to maturity from one-month T-bills over the period January 1953 to April 1974 taken from Solomon Brothers' publication "Yields and Yield Spreads". This period was chosen because there was a substantial change in the market basket used to compute the consumer price index in 1953, and this index was the one we used to measure inflation rates.

Table 1 exhibits the results of applying the Hinich time series tests to the above hypotheses. As can be observed, the first differences in the real rate, the nominal rate series, and the monthly inflation rate series all show nonlinear and non-Gaussian structure. The increments in the nominal rate series shows a possibly non-Gaussian but possibly linear structure (as might occur with a non-Gaussian random walk hypothesis), while the changes in the monthly inflation series is consistent with a Gaussian (hence linear) time series. Thus, modeling of changes in the nominal interest rates, and changes in the inflation rate by such processes as ARMA time series cannot be rejected by bispectral analysis techniques. Knowing that such processes can be adequately modeled in such a fashion is an important step foward.

Table 1. *Linearity and Gaussianity Test for the Relationship Between Real Rates r_t, Nominal Rates R_t, and Inflation Rates Δ_t*

	REJECT LINEARITY	*Z VALUE*	*REJECT NORMALITY*	*Z VALUE*
$r_t - r_{t-1} = \epsilon_t$	Yes	4.00	Yes	4.69
$\Delta_t = \alpha + \beta R_t$	Yes	17.84	Yes	21.58
$\Delta_{t-1} = \alpha + \beta R_t + \epsilon_t$	Yes	4.97	Yes	7.48
$r_t = \alpha + \beta R_t + \epsilon_t$	Yes	4.45	Yes	7.55
Δ_t	Yes	17.81	Yes	15.23
$\Delta_t - \Delta_{t-1} = \epsilon_t$ 1953-74	No	2.12	Yes	5.12
$\Delta_t - \Delta_{t-1} = \epsilon_t$ 1960-81	No	−0.27	No	2.13
$\Delta_t - \Delta_{t-1} = \epsilon_t$ 1971-81	No	−0.15	No	0.15
R_t	Yes	117.39	Yes	46.16
$R_t - R_{t-1} = \epsilon_t$	No	1.17	Yes	3.59
$R_{t-1} = \alpha + \beta R_{t-2} + \epsilon_t$	Yes	4.82	Yes	7.98

REFERENCES

Bellhouse, D. R., and H. H. Panjer (1981), "Stochastic modeling of interest rates with applications to life contingencies—Part II." *The Journal of Risk and Insurance* **48**, 628–637.

Box, G., and G. Jenkins (1966), *Time Series Analysis, Forecasting and Control.* San Francisco: Holden-Day.

Boyle, P. (1978), "Immunization under stochastic models of the term structure." *Journal of the Institute of Actuaries* **105**, 177–187.

Brillinger, D. (1975), *Time Series, Data Analysis and Theory.* New York: Holt, Rinehart and Winston.

Cox, J. C., J. Ingersoll, and S. Ross (1977), "A theory of term structure of interest rates and the valuation of interest dependent claims", working paper.

Fama, E. F. (1975), "Short-term interest rates as predictors of inflation." *American Economic Review* **65**, 269–282.

Granger, C., and O. Morgenstern (1963), "Spectral analysis of New York stock market prices." *Kyklos* **16**, 1–27.

Hinich, M. (1982), "Testing for Gaussianity and linearity of a stationary time series." *Journal of Time Series Analysis* **3**, 169–176.

Hinich, M., and D. Patterson (1985), "Evidence of nonlinearity in daily stock returns." *Journal of Business and Economic Statistics* **3**, 69–77.

Panjer, H. H., and D. R. Bellhouse (1980), "Stochastic modeling of interest rates with applications to life contingencies." *The Journal of Risk and Insurance* **47**, 91–110.

Patterson, D. (1983), "BISPEC: a program to estimate the bispectrum of a stationary time series." *American Statistician* **37**, 323–324.

Priestley, M. (1981), *Spectral Analysis and Time Series*, Volume 2. New York: Academic Press.

Rosenblatt, M. (1983), "Cumulants and cumulant spectra." In *Handbook of Statistics*, Vol. 3, ed. D. Brillinger and P. Krishnaiah, pp. 369–382. Amsterdam: North-Holland.

Subba Rao, T. (1983), "The bispectral analysis of nonlinear stationary time series with reference to bilinear and time series models." In *Handbook of Statistics*, Vol. 3, ed. D. Brillinger and P. Krishnaiah, pp. 293–319. Amsterdam: North-Holland.

Subba Rao, T., and M. Gabr (1980), "A test for linearity of stationary time series." *Journal of Time Series Analysis* **1**, 145–158.

Subba Rao, T., and M. Gabr (1984), *Introduction to Bispectral Analysis and Bilinear Time Series*. Berlin: Springer-Verlag.

Phelim P. Boyle [1]

PERSPECTIVES ON MORTGAGE DEFAULT INSURANCE

ABSTRACT

Mortgage default insurance differs from most of the other types of insurance underwritten by insurance companies. In particular, the risk units covered by a mortgage insurer are often highly correlated since some of the determinants of default are related to either economy-wide or regional factors. In such situations, the risk cannot be diversified away by underwriting more risk units. It is argued in the present paper that traditional actuarial approaches to the pricing and risk management of this type of business are inadequate. Some new paradigms from the field of financial economics are used to provide alternative approaches to these problems. The paper starts with a brief description of mortgage loan default insurance and describes some aspects of the current situation in Canada and the United States. Next, there is a discussion of informational asymmetry in the context of the mortgage default insurance market. This provides a useful framework for discussing the relationships between risk sharing and incentives. This is followed by a discussion of the determinants of default risk. One approach to the pricing of this insurance is to use an option pricing analysis. Some numerical illustrations are given to illustrate the application of this type of approach.

1. INTRODUCTION

The purpose of this paper is to examine certain aspects of mortgage loan default insurance. This branch of insurance is not very well known but it is of considerable current interest in both Canada and the United States. A mortgage loan default insurance policy protects the mortgage lender in the event that the homeowner defaults. This type of insurance was first introduced in

[1] Accounting Group, Faculty of Arts, University of Waterloo, Waterloo, Ontario, N2L 3G1

I. B. MacNeill and G. J. Umphrey (eds.), Actuarial Science, 185–199.

Canada in 1954 by the Federal Government. A comprehensive introduction to the development of mortgage loan default insurance is provided by Woodward (1959). Currently there are two insurers of this business operating in Canada: one private and one public. Mortgage loan default insurance was introduced in the United States in 1885, but the private mortgage insurance industry was virtually wiped out during The Great Depression. The Arthur D. Little (1975) study on mortgage loan default insurance provides useful information on the development of mortgage insurance in the United States as well as on various institutional aspects of this market. In 1934 the United States Government began to provide public mortgage loan default insurance, and private insurance recommenced in 1956 in the state of Wisconsin. There are currently a number of private insurers underwriting this type of business in the United States.

It is claimed in the present paper that traditional actuarial approaches to mortgage insurance are inadequate. Under conventional insurance contracts the risks can normally be assumed independent. Under mortgage insurance the claims arise from both independent and correlated factors. When risks are correlated, assembling large numbers of them into a portfolio does not lead to a reduction in the premium loading. Furthermore, the existence of correlated risks raises difficult problems concerning the appropriate reserving and investment strategies to be adopted. Some recent developments in financial economics can be of assistance here. In particular, the value of the insurance contract can be computed using an option pricing framework.

Another area where the traditional actuarial approach can benefit from some cross fertilization with current economic theory is the area of contract design and incentive provision. The structure of an insurance contract and the incentives it contains can have a profound impact on the claims incurred. Much of the conventional actuarial literature views the claims generating process as a mechanical stochastic process. Recently, considerable progress has been made in modelling the interplay between risk sharing and incentive provision in the design of optimal contracts. It is suggested that the insights from this research can be helpful in examining insurance contracts.

The layout of the paper is as follows: Section 2 provides a description of mortgage default insurance and traces its evolution in Canada. Section 3 provides an agency theory perspective and argues that the traditional form of the mortgage default insurance contract does not incorporate efficient incentives. Section 4 examines the empirical evidence on the determinants of default risk. In Section 5, the relationship between the default risk on a traded security and its yield to maturity is analysed and it is suggested that the default risk could be handled in the same way as the default risk in corporate bonds. In other words, riskier mortgages would carry a higher interest rate. An option pricing framework to determine fair mortgage insurance premiums is introduced in Section 6. It is shown that there are wide premium variations across different initial loan to value ratios and also across different mortgage instruments. The final section includes some concluding remarks.

2. MORTGAGE INSURANCE: INSTITUTIONAL FEATURES

This section describes mortgage insurance in Canada. Consider a mortgage agreement between a lender and a homeowner. Assume the initial amount of the mortgage is M and it is being amortised at rate of interest R per annum (p.a.) over N years. If the lender is concerned about the probability of default on the loan or if the loan is a high ratio one (where M exceeds 75% of the house value), mortgage insurance may be required. The insurance contract is between the lender and the mortgage insurer. The premium for the contract, P, is paid by the homeowner and is added to the principal so that the total amount of $(M + P)$ will be amortised over N years. The risk of default is shifted from the lender to the mortgage insurer. The presumed benefit to the homeowner is that he obtains the mortgage on terms that would be unavailable without the insurance.

In Canada, public mortgage insurance is available from The Canada Mortgage and Housing Corporation (CMHC). CMHC effectively sets the premium rates for this type of insurance and until 1984 the premium for a low ratio loan was 1% of the mortgage loan and the premium for a high ratio loan was $1\frac{1}{2}$% of the basic mortgage loan. These premium levels also represented the maximum rates charged by private insurers and so private insurers have had to compete on other dimensions such as service. As of 1985 there was one remaining private insurer operating in Canada: The Mortgage Insurance Company of Canada (MICC). In the United States premium levels have been historically higher than in Canada and tend to be on an annual basis.

If a claim arises under a CMHC insurance contract, the lender secures and conveys a clear title to CMHC and receives a lump sum payment covering the principal plus accrued interest and other expenses, such as legal costs and utility bills. A lender also has the option of retaining the property and selling it on the open market. As can be imagined, lenders tend to retain the more attractive properties. In contrast, MICC has more scope in settling claims. MICC can elect to pay 25% of the principal amount owing plus all expenses and retain the property for subsequent sale.

The structure of the mortgage insurance contract has remained relatively unchanged during the past 30 years. The contract does not provide for meaningful risk sharing nor does it provide appropriate incentives. For example, a homeowner in default has little incentive to maintain the property, whereas such incentives would benefit the mortgage insurer. Premiums do not vary across a wide range of risk classes resulting in considerable risk subsidies. The premium rates have tended to remain fixed during very long periods.

3. AN AGENCY THEORY PERSPECTIVE

As indicated earlier, economic theorists have made considerable progress in constructing models involving risk sharing and incentives under asymmetric information. These models have provided useful insights into the structure of optimal contracts.

In general, two parties will agree to form a contract or enter into a relationship to reduce risk if both parties stand to gain. The welfare gain can be measured in terms of utility. In an insurance setting it is rare for both the insurer and the insured to have symmetric information. In many cases of interest, the events which trigger a claim under a policy may be partly random and partly under the control of the insured. It will normally be optimal to structure the contract so that the private actions of one party which improve the other party's welfare are rewarded and that penalties are imposed for unfavourable actions. Given certain assumptions about the individuals' preferences and the distribution of the outcome, it is possible to characterize the optimal contract. The provision of appropriate incentives interferes with the pure risk-sharing aspect of such contracts but, nevertheless, it is possible to derive the 'best' contract for a given set of circumstances. Mirless (1976) and Holmstrom (1979) have made seminal contributions to this area.

Within a pure risk-sharing framework, the standard fixed rate mortgage appears to be a poorly designed instrument. Assume that the homeowner is risk averse and that the lending financial institution is risk neutral. Under these circumstances, the optimal contract in the case of pure risk sharing would involve the risk neutral party assuming all the price risk. This is exactly the opposite of what happens in practice. However it must be remembered that the homeowner derives consumption value from the ownership of his property. If we also allow for moral hazard in the sense that the house value depends in part on the level of care bestowed on it by the homeowner, then the optimal contract involves some risk sharing. Mark Wolfson (1985) analysed this type of situation and concluded that the optimal contract should involve both parties sharing in changes in the value of the house.

The mortgage default insurance contract is layered on top of the mortgage itself and a complete analysis of the optimal contract would involve both the mortgage instrument itself and the mortgage default insurance contract. Such situations involving three parties are very difficult to model. However, related work using simpler models would suggest that the mortgage insurance contract design is suboptimal. The conventional contract does not provide for efficient risk sharing. For example, if there were a deductible and the lender had to absorb some of the default losses, then this would provide the lender incentives to be more careful in underwriting and claims settlement. The lender has usually more contact with and influence on the borrower than the mortgage insurer.

The homeowner has little incentive to maintain the property once default seems likely. In most cases, the homeowner will have low current income at this time and, by prolonging foreclosure, he has the opportunity to reside rent-free in the house. In these circumstances, he is likely to expend minimal effort on care and maintenance. If the contract were to provide some inducement or bonus for good stewardship in these circumstances, this would tend to result in the defaulted houses having a higher market value since a run-down house is much less attractive in the real estate market.

4. DETERMINANTS OF DEFAULT RISK

A number of research studies have been carried out to investigate the determinants of default risk. Virtually all these studies have been carried out using United States data. A survey and brief critical analysis of some of the more recent research has been given by Campbell and Dietrich (1983). Campbell and Dietrich concluded that default incidence is related to current as well as original loan/value ratios. Default incidence is also related to changes in regional unemployment rates. They also concluded that their results based on the default experience of the 1960's and 1970's provide little basis for extrapolating the future default incidence of alternative mortgages in economies with more volatile inflation and other economic variables.

Swan (1982), in a helpful paper on pricing mortgage insurance, divides the risk factors into what he labels micro and macro factors. Micro factors include borrower, loan and maturity characteristics of an individual loan. Examples of such factors would be gross debt service ratio, loan to value ratio, age of borrower, junior financing and type of loan. Macro factors would include items such as interest rate levels, inflation rates and business cycles. As Swan notes, it should be possible to diversify away the risks arising from micro risks by underwriting a spread of risks. This technique will not be able to remove macro risks.

As far as the actual default decision of borrowers is concerned, two competing hypotheses have been discussed (see Jackson and Kaserman, 1980).

The first, the equity theory of default, is that borrowers default when the gains from default exceed the costs. The second is the ability-to-pay theory which suggests that borrowers will not default as long as they have sufficient income to meet their periodic mortgage payments. In practice, there may be an interaction of both of these influencing the default decision. Consider one scenario. A homeowner becomes unemployed. He continues making mortgage payments as long as he can. Then, having exhausted all income sources, he considers using the equity in his home to secure additional funds. If an appraisal indicates that the value of his home lies below the value of the outstanding mortgage, the incentive to default increases.

In discussing the factors which influence default risk and claim size, it seems ironic that, in recent years, government actions have sometimes represented a significant factor. The law in Alberta was amended so that lenders no longer have the covenant of the borrower as additional security on a mortgage. In these circumstances, there is a strong temptation for borrowers with the financial capacity to pay, to default. At the Federal level, the ill-fated foam-insulation programme severely reduced certain house values and increased the incentive for borrowers to default.

Another important factor which has a bearing on the default incidence is the design of the mortgage instrument. In recent years, because of increased interest rate volatility, long term fixed interest contracts such as traditional whole life insurance and long term mortgages have become less attractive. The term of mortgage loans has shortened. In addition, high and volatile inflation rates have given rise to the so-called tilt problem in the case of a standard level payment mortgage (see Carr and Smith, 1983). Under a level payment mortgage, the real percentage of income devoted to mortgage repayments starts high and declines over the term of the loan in an inflationary era.

By the same token, the equity in the home increases fairly quickly (see Intervention and Efficiency, 1982, page 165). A variety of new mortgage designs (see Carr and Smith, 1983) have been proposed to address the tilt problem. Under these new designs, the initial gross debt service ratios are smaller and the equity build-up in the house is slower. This means that the loan to value ratio remains higher than under a level payment mortgage. Consequently, the default incidence under such instruments should be greater *ceteris paribus*. This effect will be accentuated if it is accompanied by an increase in the volatility of house prices. Statistics of default gleaned from studies of conventional mortgages need to be applied with great care to these newer instruments. It is also possible that, given a menu of different mortgage designs, borrowers with particular risk characteristics and attributes may select a contract design that is very good value for them. This phenomenon is known in the insurance literature as adverse selection.

For example, buyers of annuity contracts have longer life expectancies than the population at large. Actuaries use much heavier mortality tables to compute premiums for life insurance than for annuity contracts.

Adverse selection may also distort the expected relationship between certain risk attributes and default incidence. For example, the higher the initial loan to value ratio the greater is the probability of default other things being equal. This is a result obtained in particular from the option pricing approach discussed in Section 6. However, a low loan to value mortgage may be insured because the lender has reason to believe it is a risky loan. There is no mandatory requirement for loans with a loan to value ratio below 75% to be insured. The existence of the insurance may in some cases carry information. The author understands that lenders in Canada may insist on insurance of low ratio loans for other reasons depending on the supply of mortgage funds but published empirical evidence on this does not appear to be readily available.

5. INTERNALIZING THE DEFAULT RISK IN THE YIELD

In the case of many fixed interest investments, the risk of default is internalized in the yield of the security. This is the case, for example, with corporate bonds. It is possible to visualize the mortgage market operating in this manner. In this section we illustrate by means of a simple example how the default probability is incorporated in the yield.

Assume a simple contract which, at the end of T years, makes a payment of F with probability p and a payment of zero with probability $(1-p)$. Denote these events by E_1 and E_2. We have:

Event	Probability	Payment
E_1	p	F
E_2	$(1-p)$	0

Assume that the risk free interest rate is R and that the term structure of interest rates is level. Under risk neutral valuation, the market value of the contract, V, is equal to:

$$[pF + (1-p)0]\exp(-RT) = pF\exp(-RT).$$

The yield to maturity $\overline{R}$ is given by

$$V = F\exp(-\overline{R}T).$$

Hence the expression for p is

$$\exp[-(\overline{R} - R)T]. \tag{1}$$

This formula is intuitively appealing because the difference between the yield to maturity and the risk free rate represents a default risk premium and this quantity is directly related to the probability of default, p, through equation (1).

Of course, if the distribution is more complex, then the relationship between the probability of default and the yield to maturity is more intricate. In the case of mortgages, lending institutions could make their own assessment of the default probability and internalise it in the offered mortgage rate. Under this scenario there would be no need for mortgage insurance. If it were felt that the quoted rates for some special risk groups were too high, government initiatives could be used to assist these borrowers.

6. AN OPTION PRICING APPROACHES TO PREMIUM DETERMINATION

Although the option pricing model requires very strong assumptions it has a number of advantages as a method to price mortgage default insurance. In this section, we give a brief description of the model, discuss some of its limitations, and illustrate its practical usefulness.

A number of recent papers have used the option pricing framework to price mortgage default insurance. Examples include Epperson *et al.* (1985) and Foster *et al.* (1984). Within the option pricing framework, the homeowner will default whenever the market value of his house falls below the outstanding mortgage balance. If it is assumed that the default option can only be exercised at a fixed time, the insurance contract can be viewed as a European put. If the option to default can be exercised at any time then it resembles an American put option. One of the main determinants of default in this scenario is the future evolution of house prices. In a more realistic model there are a number of other relevant items that would be included, such as stochastic interest rates and transactions costs, but these are ignored here for the sake of simplicity.

To proceed, it is necessary to introduce some notation. Let

T = time to maturity of European option to default,

Initial House Price	\$ 100,000
Standard deviation of log of house price changes	20% p.a.
Riskless interest rate	12% p.a. effective
Mortgage interest rate	13% p.a. effective
Term of mortgage	25 years
Period of Amortisation	25 years
Type of Mortgage	Fixed Rate Mortgage (FRM)
Nature of Default Option	American option

$M(T)$ = mortgage outstanding at maturity of option, and

$S(T)$ = market value of mortgaged property at maturity of option.

For a European option the boundary value at maturity is:

$$\text{Maximum } [[M(T) - S(T)], 0].$$

By making certain assumptions about the stochastic movement of house prices, the current value of such a European option can be obtained. To derive the Black Scholes formula, it is assumed that the ratio of successive house prices follows a lognormal distribution. Under the Black Scholes assumptions, the European put option has a closed form solution.

In the case of an American put option, which is somewhat more realistic in this case, there is no known analytic solution. One has to solve a parabolic differential equation to value the option. Since the mortgage balance outstanding changes as payments are made, the exercise price of the option changes during its lifetime. For the moment, we assume these changes are deterministic. Let $M(0)$ denote the initial outstanding balance and assume that R is the continuously compounded interest rate used for amortisation. The principal outstanding under a level payment mortgage after 'K' years is

$$\frac{M_{(0)} a_{\overline{N-K}|}}{a_{\overline{N}|}}. \tag{2}$$

For realistic parameter values this function declines very gently in the early years and then much more steeply as K approaches N. This type of fixed rate mortgage was until recently the most common type of residential mortgage. Here we are ignoring the option to renegotiate the terms of the mortgage after a period such as 3 or 5 years.

Numerical estimates of the fair insurance premiums have been obtained using the Cox, Ross, Rubinstein (1959) binomial lattice approach. The following parameter values were used:

Table 1. *Illustrative Fair Insurance Single Premiums for Default Option (American Option) on Fixed Rate Mortgage*

Initial Loan to Value Ratio	Dollar Value of Insurance	Insurance Premium as Percentage of Initial Mortgate
		%
80%	$ 800	1.0
85%	1,310	1.5
90%	2,100	2.3
95%	2,950	3.1

Note that there is considerable variation in the insurance premium rates across different initial loan to value ratios.

It is instructive to examine the corresponding premiums in the case of some alternative mortgage instruments. The graduated payment mortgage is of interest in this connection because the option pricing approach predicts much higher insurance premiums in this case. A graduated payment mortgage (GPM) is one where the mortgage payments increase each year by a certain percentage. Let R denote the annual interest rate on a corresponding fixed rate mortgage and g denote the growth factor under the GPM. If $M(0)$ is the initial mortgage balance the initial payment is

$$\frac{M(o)}{a_{\overline{n}|}} = L, \tag{3}$$

where the annuity is calculated at rate R and N is the amortisation period.

The payment at the end of year t, is equal to:

$$L(1+g)^t.$$

The outstanding balance at the end of year t, $M(t)$, can be computed recursively:

$$M(t) = M(t-1) + RM(t-1) - L(1+g)^t. \tag{4}$$

Under a GPM, the outstanding mortgage balance increases for a period and then decreases so that the loan is amortised at the end of the period. Table 2 compares the time paths of the mortgage outstanding under both an FRM and a GPM. The underlying parameters are:

$$M(0) = 80,000$$
$$R = 13\% \text{ p.a.}$$
$$g = 5\% \text{ p.a.}$$
$$N = 25 \text{ years}$$

Table 2. *Principal Outstanding Under FRM and GPM*

t	FRM	GPM
	$	$
0	80,000	80,000
5	76,670	93,510
10	70,530	103,510
15	59,220	102,920
20	38,390	77,600
25	-----	-----

The GPM balance rises to a maximum of about $105,000 after 12 years before starting to decline. Recall that the put option value is equal to the maximum of the excess of the mortgage principal and the house price and zero. It is clear that for two identical house price trajectories the GPM will give rise to more claims under mortgage insurance. As the numbers in Table 3 indicate, the GPM default risk is very substantially greater than that for the FRM. When GPMs were first issued this was not adequately recognized by insurers. Note that the mortgage premiums under the GPM are approximately double those under the FRM.
In recent years there have been different types of new mortgage instruments such as Variable Rate Mortgages, Price Level Adjusted Mortgages and Shared Appreciation Mortgages. The option pricing framework provides a mechanism for estimating insurance premiums under these contracts.

The assumptions required to implement the option pricing approach are strong. For example, it assumes that the sole determinant of mortgage defaults is the homeowners equity in the property whereas we saw in Section

Table 3. *Illustrative Single Premiums for Mortgage Default Insurance. Parameter Values as in Table 1*

Initial Loan to Value Ratio %	Dollar Value of Insurance	GPM Insurance Premium as Percentage of Mortgage	FRM Insurance Premium as Percentage of Mortgage
	$		
80	1976	2.5	1.0
85	2899	3.4	1.5
90	4017	4.5	2.3
95	5439	5.7	3.1

4 that a number of other variables influence the default decision. To date there has been a tendency to assume rather simplistic boundary conditions. Most of the work to date has used the lognormal diffusion assumption for house prices and it is unlikely that this is the most appropriate stochastic process. The variance of the series of house price returns is difficult to estimate. Hoag's (1980) research indicates that the appraisal process introduces a degree of smoothing so that the historical variances may themselves be underestimates. There is a considerable range in the variance estimates across time and regions as Table 4 indicates. In spite of its limitations the option pricing approach seems to offer certain advantages in this area. It produces market based estimates of insurance premiums. The option based approach can provide a useful analytical framework in a way which is impossible under the traditional actuarial approach. Furthermore, it can be used to estimate mortgage default premium rates for new types of mortgage design.

7. CONCLUSION

This paper has examined a number of aspects of mortgage insurance. Drawing mainly on the Canadian experience, the evolution of mortgage insurance was examined. It was alleged that the current financial problems stem from inadequate premiums and inefficient contract design. The rigidi-

Table 4. *Historical Estimates of the Standard Deviations of Residential House Prices*

Location	Time Period	Standard Deviation	Source
		% p.a.	
U.S.A.	1960–1980	9	Foster & Van Order (1984)
Southern California	1960–1967	11.9	Assay (1978)
Canada	1974–1982		Present Author (based on Royal Trust Survey)
Mississauga (Toronto)	"	8.1	"
Mount Royal (Montreal)	"	11.7	"
Richmond (Vancouver)	"	21.0	"
Calgary	"	12.4	"

ties imposed by CMHC on the insurance market-place have made it difficult for MICC to operate on a financially sound basis.

A basic thrust of this paper is that the design and actuarial techniques used in the case of mortgage insurance are somewhat outdated. Two useful paradigms from modern financial economics were introduced and shown to provide useful insights. The problems caused by asymmetric information for optimal risk sharing were discussed in this context. Modern agency theory provides a framework for characterising optimal contracts in terms of their risk sharing and incentive aspects. While a complete analytical solution would be hard to obtain, in the present case, useful insights were obtained. In the case of premium computations it was shown that the option pricing framework provided a useful approach. Numerical estimates were obtained for the insurance premiums for both fixed rate mortgages and graduated payment mortgages. This approach indicates clearly the much higher risks that the insurer faces under the GPM.

Because of space limitations, the paper has had to restrict attention to only a few aspects of mortgage insurance and, even here, the treatment has sometimes been general. Important topics not dealt with include financial risk management for mortgage insurers, the role of public and private insurers and the question of reinsurance.

ACKNOWLEDGMENTS

The author is grateful to the Natural Sciences and Engineering Research Council of Canada for research support. He is also grateful to Dennis Myette of the Canada Mortgage and Housing Corporation for stimulating his interest in this area. All views expressed in this paper are attributable to the author.

REFERENCES

Asay, M. R. (1978), "Rational mortgage pricing". Unpublished manuscript, December 1978.

Assessment Report, Mortgage Loan Insurance (1984) CMHC, Program Evaluation Division.

Campbell, T. S., and J. K. Dietrich (1983), "The determinants of default on insured conventional residential mortgage loans". *Journal of Finance* **5**, 1569–1581.

Carr, J. L., and L. B. Smith (1983), "Inflation, uncertainty and future mortgage instruments". In *North American Housing Markets into the Twenty-first Century*, ed. G. Gau and M. Goldberg, pp. 203–231. Cambridge MA: Ballinger.

Cox, J., S. A. Ross, and M. Rubinstein (1979), "Option pricing, a simplified approach". *Journal of Financial Economics* **7**, 229–263.

Epperson, J. F., J. B. Kan, D. C. Keenan, and W. J. Muller (1985), "Pricing default risk on mortgages". *Journal of the American Real Estate and Urban Economics Association* **13**, 261–272.

Foster, C., and R. Van Order (1984), "An option-based model of mortgage default." *Housing Finance Review* **3**, 351–372.

Hoag, J. (1980), "Towards indices of real estate value and return". *Journal of Finance* **35**, 569–580.

Holmstrom, B. (1979), "Moral hazard and observability". *Bell Journal of Economics* **10**, 74–91.

Intervention and Efficiency, 1982, Economic Council of Canada.
Jackson, J. R., and D. Kaserman (1980). "Default risk on home mortgage loans: a test of competing hypotheses." *Journal of Risk and Insurance* **47**, 678–690.
Little, Arthur D. (1975) Inc., "The private mortgage insurance industry". Research report, Arthur D. Little, Inc., 35 Acorn Park, Cambridge, MA 02140.
Smith, C. (1980), "On the theory of financial contracting: the personal loan market". *Journal of Monetary Economics* **6**, 333–358.
Swan, C. (1982), "Pricing private mortgage insurance". *Journal of the American Real Estate and Urban Economics Association* **10**, 276–296.
Wolfson, M. (1985), "Tax incentive and risk-sharing issues in the allocation of property rights: the generalized lease-or-buy problem". *Journal of Business* **58**, 159–171.
Woodward, H. (1959), *Canadian Mortgages*. Don Mills: Collins.

Keith P. Sharp [1]

TIME SERIES ANALYSIS OF MORTGAGE RATE INSURANCE

ABSTRACT

A Mortgage Rate Protection Plan was introduced effective March 1, 1984 by the federal government of Canada. Under the plan, mortgage borrowers could, in exchange for payment of a single premium, insure themselves against the risk of large increases in their monthly mortgage payment. It is shown in this paper that the SARIMA model

$$(1-\Phi_1 B^{12})(1-\phi_1 B)(1-B)z_t = a_t$$

satisfactorily represents monthly interest rate data z_t. An expression for the appropriate insurance premium is derived, and comparisons are made with the premium actually being charged by CMHC.

1. INTRODUCTION

At inception of a residential mortgage in Canada, the borrower may choose the term, usually 1, 2, 3, 5 or 7 years, for which the interest rate will be fixed. At renewal, the interest rate changes to the then current market rate. During the early 1980's, many persons renewed mortgages at unexpectedly high rates, and pressure arose for government action to control the risks arising from volatile interest rates. As a result, the Mortgage Rate Protection Plan (MRPP) was established effective March 1, 1984 by the Canada Mortgage and Housing Corporation (CMHC). (The program is described in the pamphlet "Mortgage Rate Protection Program", document number NGA 5813 84/08, CMHC, Ottawa.)

At mortgage inception or renewal, borrowers may choose to participate in the MRPP by paying a single premium of $1\frac{1}{2}\%$ of principal, regardless

[1] Department of Statistics and Actuarial Science, University of Waterloo, Waterloo, Ontario N2L 3G1

I. B. MacNeill and G. J. Umphrey (eds.), Actuarial Science, 201–218.

of the inter-renewal period. If interest rates at renewal have risen by more than a 2% deductible, then the MRPP pays 75% of the excess of the market monthly payment over the monthly payment calculated at the original rate plus 2% per annum. Any portion of the rate rise above 12% per annum does not receive a subsidy. The period of subsidy is equal to the inter-renewal term chosen at the time of insurance purchase, even if the subsequent renewal is for some other term. For example, at inception of a $70,000 mortgage with amortization over 25 years and an initial interest rate of 12.5% per annum compounded semi-annually, the monthly payment is $747. If at renewal after 5 years market rates are 17.5% per annum, then the full market payment of $978 per month is reduced by MRPP payments of $105 per month. Here the $105 is calculated as 75% of the difference between $978 and the monthly payment calculated at 14.5% per annum, $838.

Under the MRPP, the principal outstanding at the end of the subsidised period is calculated at the market rate applicable to the subsidised period, so is not affected by the existence of the insurance. The MRPP insurance can be purchased at the renewal subsequent to the initial purchase, but the basic interest rate is the market rate at the renewal. Thus the mortgage borrower is not fully protected against increases in the monthly payment resulting from a high market rate two inter-renewal periods after the initial insurance purchase.

The focus of this paper is a check on the appropriateness of the $1\frac{1}{2}\%$ single premium charged by the CMHC. Section 2 reviews SARIMA time series methods. Those familiar with SARIMA techniques may prefer to proceed directly to Section 3 in which the time series of monthly mortgage interest rates is modelled. In Section 4, formulas are derived whereby the model parameters found for the interest rate series can be used to calculate the expected value of the payout under the insurance. Section 5 includes comments on the results and on the MRPP as it currently exists.

2. SARIMA MODELLING METHODS

The books by Box and Jenkins (1978) and by Abraham and Ledolter (1983) have popularised the use of autoregressive integrated moving average (ARIMA) models. The models have found applications in many fields, but particularly in the analysis and short-term forecasting of economic and business series.

If the series to be analysed is denoted by w_t, then the first order autoregressive model, denoted by AR(1) or ARIMA(1,0,0), is given by

$$w_t = \phi_1 w_{t-1} + a_t. \tag{1}$$

Here ϕ_1 is a parameter which must be estimated, and $a_1, a_2, a_3, \ldots$ is a sequence of independent identically distributed "shock" terms, with the normal distribution

$$a_t \sim N(0, \sigma^2). \tag{2}$$

Using B, the back shift operator, the AR(1) model (1) can be re-expressed as

$$(1 - \phi_1 B)w_t = a_t. \tag{3}$$

Another variation can be derived by modelling the first differences of the original data series z_t,

$$w_t = z_t - z_{t-1} = (1 - B)z_t. \tag{4}$$

Then the first order autoregressive model for the first difference is denoted ARIMA$(1, 0, 0)$ and is written

$$(1 - \phi_1 B)(1 - B)z_t = a_t. \tag{5}$$

The moving average family models the data series as a weighted sum of past shocks. For instance, the MA(1) or ARIMA$(0, 0, 1)$ model is given by

$$w_t = (1 - \theta_1 B)a_t. \tag{6}$$

The ARIMA(p, d, q) model is given by

$$\phi(B)(1 - B)^d z_t = \theta(b)a_t, \tag{7}$$

where

$$\phi(B) = 1 - \phi_1 B - \phi_2 B^2 - \ldots - \phi_p B^p \tag{8}$$

and

$$\theta(B) = 1 - \theta_1 B - \theta_2 B^2 - \ldots - \theta_q B^q. \tag{9}$$

Many series exhibit seasonal fluctuations. In order to model these series parsimoniously (i.e., with the minimum number of parameters ϕ, θ, etc. to be estimated) multiplicative seasonal models are useful. The form is

$$\Phi(B^s)\phi(B)(1 - B^s)^{D_s}(1 - B)^d z_t = \theta_0 + \Theta(B^s)\theta(B)a_t, \tag{10}$$

where s is the seasonality (e.g., 12 for a monthly series),

$$\Phi(B^s) = 1 - \Phi_1 B^s - \Phi_2 B^{2s} - \ldots - \Phi_{P_s} B^{P_s s} \tag{11}$$

and

$$\Theta(B^s) = 1 - \Theta_1 B^s - \Theta_2 B^{2s} - \ldots - \Theta_{Q_s} B^{Q_s s}. \tag{12}$$

The above model is denoted SARIMA$(p, d, q)(P_s, D_s, Q_s)_s$. A constant term θ_0 is included for generality, but often θ_0 is found to be zero.

It can be seen that the SARIMA family of models is very rich. It is possible to adequately model most economic series, provided that the series does not dramatically change character or behavior during the modelled period. For instance, it would be difficult to adequately model monthly world oil prices over a period which included the early 1970's. Although SARIMA models can contain many parameters θ_i, ϕ_i, Θ_i and Φ_i which must be estimated from the data, the typical series can be modelled parsimoniously. Often a series can be well represented by a model with only two or three such parameters.

Forecasts can be provided by SARIMA models. These can be of considerable practical value in, for instance, the case of a series for sales of a steadily expanding corporation in a seasonally dependent business. For series such as the Consumer Price Index or stock prices, forecasts tend not to be very useful. The confidence interval for forecasts with a lead time of even 3 or 4 periods is so large as to render the forecast redundant. However, SARIMA analysis does enable one to make an accurate estimate of the volatility of a series. This is of value in determining the appropriate premium to be charged for mortgage rate insurance.

Box and Jenkins (1978) and Abraham and Ledolter (1983) described an iterative process whereby SARIMA models can be derived for a given series. Model building is a three stage process:

(1) identification,
(2) estimation,
(3) testing.

If the model fails stage 3 (testing), then a new model is identified and tested.

The identification stage consists of four steps:

(a) choice of transformation of raw data (e.g., logarithm of raw data) which stabilizes the variance,
(b) choice of degree of differencing (d and D_s) which leads to a stationary series, i.e., one without a change in mean or variance during the period modelled,
(c) choice of orders of model, p, q, P_s and Q_s,
(d) decision as to whether the constant term θ_0 in (10) is significantly different from zero.

The identification stage is accomplished through analysis of plots of the autocorrelation function r_k, where

$$r_k = \frac{\sum_{t=1}^{n-k}(w_{t+k} - \bar{w})(w_t - \bar{w})}{\sum_{t=1}^{n}(w_t - \bar{w})^2} \tag{13}$$

and n is the number of data points considered. Other aids in the analysis are the residual variance, plots of the differenced series, and the partial autocorrelation function (Abraham and Ledolter, 1983). The partial autocorrelation function ϕ_{kk} indicates the degree of correlation for a given lag after removal of the effects of joint correlation with intervening lags. It corresponds to the coefficient of the kth variable in a multiple regression which includes lags up to the kth:

$$w_t = \phi_{k1}w_{t-1} + \phi_{k2}w_{t-2} + \ldots + \phi_{kk}w_{t-k} + a_t. \tag{14}$$

After the parameters d, D_s, p, q, Q_s, and P_s are identified by the investigator, the parameters ϕ_i $(1 \le i \le p)$, θ_i $(1 \le i \le q)$, Φ_i $(1 \le i \le P_s)$ and Θ_i $(1 \le i \le Q_s)$ are estimated by the method of maximum likelihood. This stage is performed here through use of the TS computer package (McLeod, 1978).

The testing stage is the most important and involves the use of judgement by the investigator. Among the aids in testing are:

(a) Plots of the residuals from the model, i.e. the amount of variance accounted for by the current time random shock a_t. The plot should show only white noise.

(b) Plots of sample autocorrelation $r_{\hat{a}}(k)$ of the residuals. The autocorrelations of the residuals should be zero if the model is correct. Sample autocorrelations are nonzero, but approximately 95% of them should be within $\pm 2\sigma$ of zero. The $\pm 2\sigma$ limits are indicated on the plots of the residual autocorrelations; see Figures 3 and 4.

(c) Values of the portmanteau statistic

$$Q(K) = n(n+2)\sum_{k=1}^{K} \frac{r_{\hat{a}}^2(k)}{n-k}. \tag{15}$$

Under the hypothesis that the method is correct, Q is distributed as a chi-squared random variable with $K - P_s - Q_s - p - q$ degrees of freedom.

(d) The proportion of the variance σ^2 of the differenced data which is accounted for by the model.

3. SARIMA MODELLING OF MORTGAGE RATES

Monthly mortgage rates are given by CANSIM B14024 (Table 1) for the period January 1951 to December 1984. The data are heterogeneous because

– until 1969, rates were for mortgages with rates fixed for 25 years,

Table 1. *Cansim B14024 Mortgage Rates Jan. 1951–Dec. 1984*

1951	5.00	5.00	5.00	5.25	5.50	5.50	5.62	5.62	5.75	5.75	5.75	5.75
1952	5.70	5.70	5.70	5.70	5.80	5.80	5.85	5.85	5.75	5.80	5.80	5.80
1953	5.90	5.90	5.90	5.90	5.90	5.95	5.95	5.95	6.05	6.05	6.10	6.10
1954	6.05	6.05	6.05	6.00	6.00	6.00	6.00	6.00	6.00	6.00	6.00	6.00
1955	6.00	6.00	6.00	6.00	5.75	5.75	5.75	5.70	5.80	5.90	5.95	5.95
1956	5.95	5.95	6.00	6.00	6.00	6.05	6.15	6.35	6.40	6.55	6.65	6.65
1957	6.70	6.75	6.75	6.75	6.75	6.85	6.85	6.90	7.00	7.00	7.00	6.95
1958	6.95	6.90	6.80	6.75	6.75	6.75	6.75	6.75	6.75	6.80	6.80	6.80
1959	6.85	6.85	6.85	6.80	6.80	6.85	6.85	6.95	7.20	7.20	7.25	7.25
1960	7.30	7.30	7.30	7.30	7.25	7.25	7.15	7.15	7.10	7.00	7.00	7.00
1961	7.00	7.00	7.00	7.00	7.00	7.00	7.00	7.00	7.00	7.00	7.00	7.00
1962	7.00	7.00	7.00	6.90	6.80	6.95	7.00	7.00	7.00	7.00	7.00	7.00
1963	7.00	7.00	7.00	6.94	6.91	6.91	6.91	7.00	7.00	7.00	7.00	7.00
1964	7.00	7.00	7.00	6.95	6.88	6.88	6.88	7.00	7.00	7.00	7.00	7.00
1965	6.90	6.85	6.82	6.82	6.83	6.83	7.02	7.13	7.15	7.25	7.29	7.40
1966	7.38	7.45	7.46	7.48	7.51	7.57	7.68	7.80	7.84	7.87	7.91	7.95
1967	7.93	7.89	7.83	7.80	7.77	7.88	8.02	8.05	8.10	8.49	8.52	8.52
1968	8.83	8.84	8.96	9.20	9.23	9.18	9.14	9.12	9.03	9.01	9.09	9.10
1969	9.45	9.45	9.48	9.52	9.46	9.69	9.90	9.99	10.11	10.21	10.30	10.50
1970	10.58	10.54	10.58	10.60	10.58	10.53	10.38	10.40	10.36	10.35	10.28	10.16
1971	9.94	9.72	9.28	9.20	9.25	9.34	9.46	9.53	9.55	9.55	9.26	9.10
1972	9.04	8.93	8.97	9.03	9.16	9.37	9.41	9.41	9.38	9.35	9.30	9.22
1973	9.09	9.02	9.07	9.15	9.30	9.52	9.71	9.91	10.13	10.13	10.08	10.02
1974	10.02	10.01	10.04	10.70	11.26	11.37	11.60	11.85	12.05	12.05	12.00	11.88
1975	11.81	10.95	10.65	10.67	10.99	11.23	11.35	11.52	11.94	12.15	11.97	11.89
1976	11.84	11.80	11.90	12.03	11.99	11.93	11.86	11.83	11.76	11.60	11.56	11.27
1977	10.75	10.25	10.25	10.25	10.38	10.35	10.40	10.33	10.32	10.34	10.34	10.33
1978	10.32	10.31	10.33	10.42	10.43	10.32	10.31	10.31	10.67	10.93	11.25	11.53
1979	11.28	11.25	11.11	11.05	11.06	11.16	11.20	11.80	12.25	13.50	14.46	13.58
1980	13.26	13.50	14.69	16.94	13.99	12.92	13.09	13.44	14.50	14.87	15.00	15.60
1981	15.17	15.27	15.75	16.45	17.82	18.55	18.90	21.30	21.46	20.54	18.80	17.79
1982	18.21	18.97	19.41	19.28	19.11	19.10	19.22	18.72	17.49	16.02	14.79	14.34
1983	14.05	13.60	13.45	13.26	13.16	12.98	13.08	13.57	13.88	13.10	12.84	12.55
1984	12.55	12.52	12.82	13.51	14.26	14.53	14.96	14.45	13.99	13.72	13.25	12.74

- rates from 1969 are applicable to mortgages with rates fixed for 5 years, except in 1982–1983 when 5 year mortgages were unavailable and 2 and 3 year mortgages were substituted.

Casual inspection of Table 1 shows that mortgages rates were relatively stable from 1951 to 1967. From 1967 to 1979 there was significant variability in rates, and the period 1979 to 1984 has, of course, been characterised by extreme volatility which lessened towards the end of that period. The increase in volatility in 1979 is linked to that year's adoption by the United States Federal Reserve of a policy of adherence to money supply targets (Buser and Hendershott, 1984). Attempts to fit SARIMA models over the whole period 1951 to 1984 failed because of this change in the character of the series. Therefore, two more homogeneous sub-periods were chosen. In view of the change from 25 year fixed interest terms in 1969, the periods chosen were:

(a) January 1970 through December 1978,
(b) January 1979 through December 1984.

Various transformations were tested, but it was decided that none had significant advantages over the use of raw interest rates. The evident "wandering mean" of the raw data series, together with analysis of autocorrelation plots, led to the decision to model first differences of the interest rate series.

The sample autocorrelation function (Figure 1) shows peaks at lags 0, 12, 24 and 36. The sample partial autocorrelation function (Figure 2) has a value at lag 11 which comparison with the plotted $\pm 2\sigma$ lines indicates is significant. The evidence is thus strong that mortgage rates were seasonal over the 1970–1978 period considered in these plots. The peak rates of increase were during the summer months. This may be a result of increased house purchase activity during those months, or a result of a general increased demand for money at those times.

Figures 1 and 2 are characteristic of an AR(1) process in a 12 period seasonal model, together with an MA(1) or AR(1) process in the nonseasonal part. Indeed, the slowness of the decay of the oscillations in Figure 4, the autocorrelation plot, indicates that $D_s = 1$ models should be tried, but theses models were found to fail the diagnostic tests.

About 30 models were tested, but overall the best model was SARIMA $(1,1,0)(1,0,0)_{12}$:

$$(1 - \Phi_1 B^{12})(1 - \phi_1 B)(1 - B)z_t = \theta_0 + a_t. \tag{16}$$

The residual autocorrelation plots of Figures 3 and 4 are satisfactory for both periods, that is, January 1970 through December 1978 and January 1979 through December 1984. For the latter period the fit is slightly less good. However, the lags which have residual autocorrelations marginally outside the $\pm 2\sigma$ limits are 5, 15 and 20. These are not important lags.

For both time periods the mean θ_0 was insignificant. The estimated parameters with their standard errors for the SARIMA$(1,1,0)(1,0,0)_{12}$ models together with two of the diagnostic statistics are given in Table 2.

The parameters found for the two periods are substantially different. The 1979–1984 period is characterised by much higher volatility and by the submersion of the seasonal component in the statistical noise, and by a smaller proportion of the variance being accounted for by the model.

Table 3 indicates the 60 month forecasts which the models generate from origins December 1978 and December 1984. For instance, the forecasts from December 1984 for the mortgage rate in December 1987 is 13.04% $\pm$ 6.60%, where 6.60% corresponds to one standard deviation. As previously pointed out, the confidence interval using $\pm 1.96\sigma$ limits is sufficiently large that the forecast is of little use. Nonetheless, an indication is given of the

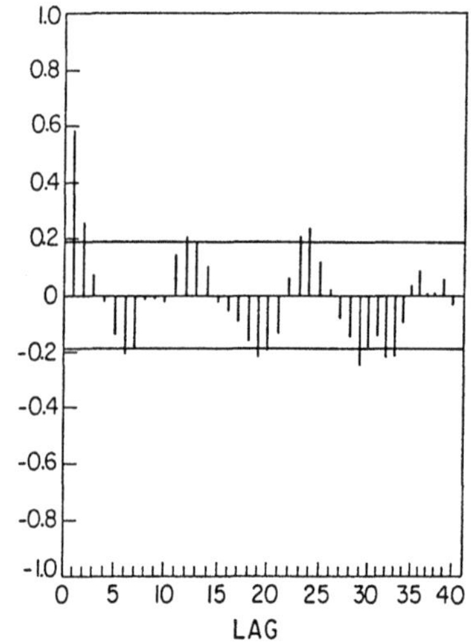

Figure 1. *Autocorrelation function of monthly changes in mortgage rates, Jan. 1970–Dec. 1978.*

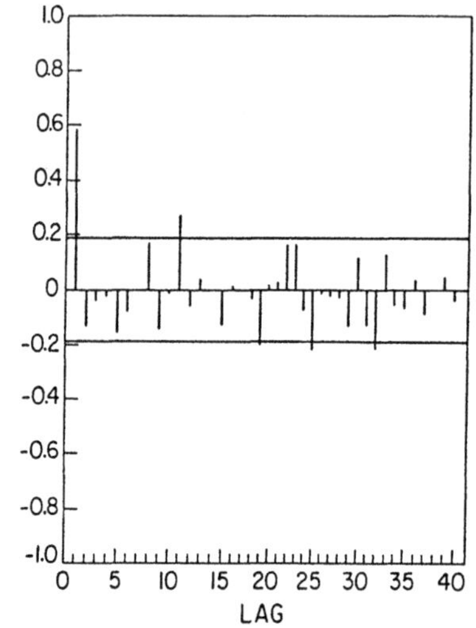

Figure 2. *Partial autocorrelation function of mortgage rates, Jan. 1970–Dec. 1978.*

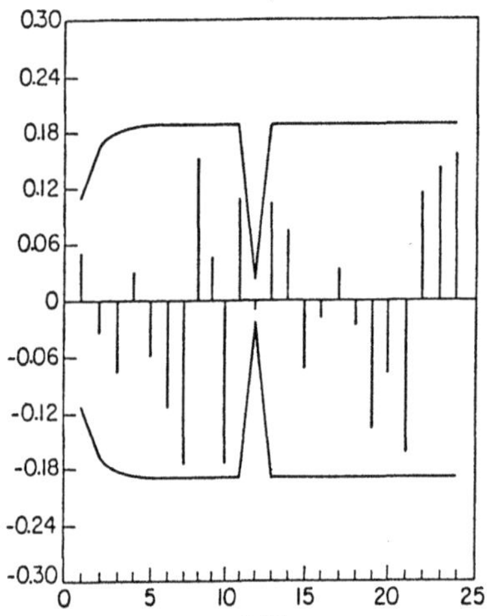

Figure 3. *Autocorrelations of residuals of SARIMA*$(1,1,0)(1,0,0)_{12}$ *model for Jan. 1970–Dec. 1978 mortgage rates.*

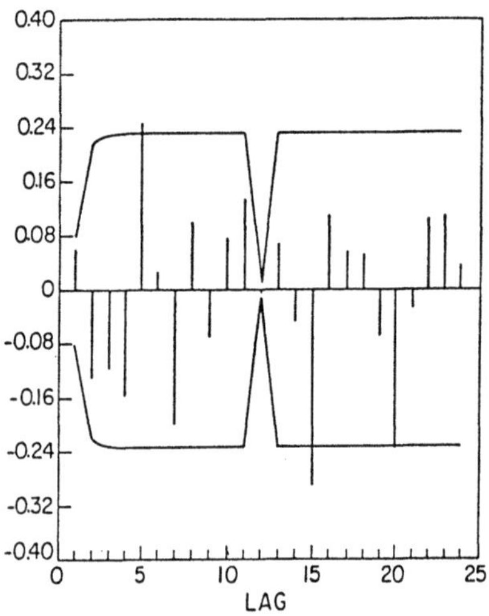

Figure 4. *Autocorrelations of residuals of SARIMA*$(1,1,0)(1,0,0)_{12}$ *model for Jan. 1979–Dec. 1984 mortgage rates.*

Table 2. *Parameters of SARIMA* $(1,1,0)(1,0,0)_{12}$ *Models*

	Jan. 1970–Dec. 1978	*Jan. 1979–Dec. 1984*
ϕ_1	0.5820±0.0786	0.3359±0.1118
Φ_1	0.1241±0.0959	-0.0509±0.1185
σ^2 (residuals)	0.02462	0.58480
σ^2 (differenced raw data)	0.03878	0.65934
$Q(24)$	32.79670	35.08526

Table 3. *Forecasts Using SARIMA* $(1,1,0)(1,0,0)_{12}$ *Models With One Standard Deviation Errors*

Lead (months)	Origin December 1978	Origin December 1984
0	11.53%	12.74%
1	11.70%±0.16%	12.57%±0.76%
2	11.80%±0.29%	12.53%±1.28%
3	11.87%±0.42%	12.51%±1.69%
12	12.19%±1.17%	12.65%±3.84%
24	12.41%±1.86%	12.85%±5.40%
36	12.57%±2.37%	13.04%±6.60%
48	12.72%±2.80%	13.22%±7.61%
60	12.88%±3.17%	13.41%±8.51%

volatility and of the minimum mean square error forecast, which indicates an increase from December 1984 levels of 12.74% to a December 1989 level of 13.41%. The effect of the higher forecast on the pricing of insurance with a 2% deductible would not be large in view of the large standard deviation of the estimate, but could be significant for other periods where the forecast can be several percent higher than the latest data point.

The satisfactory fit of the SARIMA$(1,1,0)(1,0,0)_{12}$ model, at least for the January 1970–December 1978 period, indicates that trends in interest rates do occur. The best guess at the interest rate change for the next period is related to the interest rate change in the last period (see equation (1)). This is in line with the observation that mortgage rates do appear to have periods of two or three months when they have a consistent upwards or downwards movement. After a few periods, the effect of the shocks a_t "washes out the memory" of the observed interest rate change for the last period.

The forecast of Table 3 for a December 1978 origin shows a monotonic increase towards an asymptote at large lead times of about 13%. This is a result of the modelled autoregressive nature of the first differences, and the fact that the last observation, 11.53%, shows an increase over the previous one, for November 1978, 11.25%. The increasing nature of the forecast is a mathematical expression of the intuitive view that the mortgage rates show trends, and the observed upward trend is likely to continue for a while. The standard deviation of the forecasts show a "funnel of doubt" which initially expands in a way set by the chosen model, but which for large lead times expands as the square root of the lead time in the usual "random walk" manner.

Table 3 shows also the forecasts based on January 1979 to December 1984 data. Although the most recent observations show a downward trend, the forecast after lead time one year is increasing. This is a result of the seasonal component which has a substantial effect because the data about a year before the end of the period was at a relatively low level. It will be noted that the lead 60 forecast standard deviation of 8.51% is much greater than the lead 60 forecast standard deviation of 3.17% found for the less volatile 1970–1978 period.

These forecasts serve to emphasize that the price of the mortgage insurance should theoretically depend on trends in mortgage rates up to the date of purchase. In view of the autoregressive nature of the mortgage rates, these trends do give information about the likely level of rates at renewal. However, the standard deviations of the forecasts are large, and this trend effect is not large compared with the degree of uncertainty of the forecasts. As a practical matter, insurance pricing dependent on recent rate trends would be difficult to implement.

It should be noted that the standard deviations of the forecasts depend only on the lead time and on the parameter estimates (Abraham and Ledolter, 1983). Thus they are not critically dependent on the last few observations of the period and can be regarded as being more reliable than the forecasts themselves. They can be used as a basis for pricing the insurance.

4. INSURANCE PRICING

Assume that at time $t = 0$ a mortgagor arranges to borrow (or to reborrow if a renewal is contemplated) an initial principal P_0. Amortization is over n years and the interest rate is fixed for m years. Payments are monthly in arrears. The initial interest rate is $i_0^{(2)}$ per annum convertible semi-annually. For the first m years the monthly payment is given by

$$\frac{1}{12}\frac{P_0}{a^{(12)}_{\overline{n}|}(i_0^{(2)})}. \tag{17}$$

After m years the principal outstanding is

$$P_m = P_0 \frac{a^{(12)}_{\overline{n-m}|}(i_0^{(2)})}{a^{(12)}_{\overline{n}|}(i_0^{(2)})}. \tag{18}$$

The borrower purchases mortgage rate insurance at time $t = 0$. There is an interest rate "deductible" of $i_d^{(2)}$ ($= 0.02$ for the federal program). The insurance covers increased monthly payments from time $t = m$ to $t = 2m$. Let the single premium for the insurance be denoted by I.

The present value $V(i_m^{(2)})$ at time 0 of the subsidy between time m and $2m$ applicable to a portion R of the mortgage loan depends on the time m interest rate as follows

$$V(i_m^{(2)}) = \begin{cases} 0, \quad i_m^{(2)} < i_0^{(2)} + i_d^{(2)} \\[2ex] RP_0\left(1+\frac{i_0^{(2)}}{2}\right)^{-2m} a^{(12)}_{\overline{m}|}(i_0^{(2)}) \frac{a^{(12)}_{\overline{n-m}|}(i_0^{(2)})}{a^{(12)}_{\overline{n}|}(i_0^{(2)})} \\ \qquad \times\left[\frac{1}{a^{(12)}_{\overline{n-m}|}(i_m^{(2)})} - \frac{1}{a^{(12)}_{\overline{n-m}|}(i_0^{(2)}+i_d^{(2)})}\right], \\ \qquad i_0^{(2)} + i_d^{(2)} \le i_m^{(2)} < i_0^{(2)} + i_\ell^{(2)} \\[2ex] RP_0\left(1+\frac{i_0^{(2)}}{2}\right)^{-2m} a^{(2)}_{\overline{m}|}(i_0^{(2)}) \frac{a^{(12)}_{\overline{n-m}|}(i_0^{(2)})}{a^{(12)}_{\overline{n}|}(i_0^{(2)})} \\ \qquad \times\left[\frac{1}{a^{(12)}_{\overline{n-m}|}(i_0^{(2)}+i_\ell^{(2)})} - \frac{1}{a^{(12)}_{\overline{n-m}|}(i_0^{(2)}+i_d^{(2)})}\right], \\ \qquad i_0^{(2)} + i_\ell^{(2)} \le i_m^{(2)}. \end{cases} \tag{19}$$

Then if the probability density function of $i_m^{(2)}$ conditional on data received by time 0 is given by $f(i_m^{(2)})$, the insurance value I is given by

$$I = \int_{i_0^{(2)}+i_d^{(2)}}^{i_0^{(2)}+i_l^{(2)}} V(i_m^{(2)}) f(i_m^{(2)})\, di_m^{(2)} + V(i_0^{(2)}+i_l^{(2)}) \int_{i_0^{(2)}+i_l^{(2)}}^{\infty} f(i_m^{(2)})\, di_m^{(2)}. \quad (20)$$

Equation (20) can be evaluated accurately by numerical methods. An analytic result can be found using a straight line approximation to $V(i_m^{(2)})$ based on its gradient at $i_m^{(2)} = i_0^{(2)} + i_d^{(2)}$. We use the relation

$$\begin{aligned}\frac{d}{di^{(2)}} a_{\overline{u}|}^{(12)}(i^{(2)}) &= -(1+\frac{i^{(2)}}{2})^{-1}(Ia^{(12)})_{\overline{u}|}^{(12)} \\ &= -\left(1+\frac{i^{(2)}}{2}\right)^{-1} \frac{\ddot{a}_{\overline{u}|}^{(12)} - u\left(1+\frac{i^{(2)}}{2}\right)^{-2u}}{i^{(12)}} \quad (21)\end{aligned}$$

to obtain the gradient

$$\begin{aligned}\left.\frac{dV(i_m^{(2)})}{di_m^{(2)}}\right|_{i_m^{(2)}=i_0^{(2)}+i_d^{(2)}} &= RP_0\left(1+\frac{i_0^{(2)}}{2}\right)^{-2m} \frac{a_{\overline{m}|}^{(12)}(i_0^{(2)}) a_{\overline{n-m}|}^{(12)}(i_0^{(2)})}{a_{\overline{n}|}^{(12)}(i_0^{(2)})[a_{\overline{n-m}|}^{(12)}(i_0^{(2)}+i_d^{(2)})]^2} \\ &\times \left(1+\frac{i_0^{(2)}+i_d^{(2)}}{2}\right)^{-1} (Ia^{(12)}) a_{\overline{n-m}|}^{(12)}(i_0^{(2)}+i_d^{(2)}). \quad (22)\end{aligned}$$

The forecasts of $i_m^{(2)}$ produced by SARIMA analysis are normally distributed (Abraham and Ledolter, 1983):

$$f(i_m^{(2)}) \sim N(\mu_m, \sigma_m^2), \quad (23)$$

where μ_m and σ_m are the forecast mean and standard deviation. Thus equation (20) becomes

$$\begin{aligned}I = \left.\frac{dV(i_m^{(2)})}{di_m^{(2)}}\right|_{i_0^{(2)}+i_d^{(2)}} &\int_{i_0^{(2)}+i_d^{(2)}}^{i_0^{(2)}+i_l^{(2)}} \left(i_m^{(2)} - i_0^{(2)} - i_d^{(2)}\right) \frac{1}{\sigma_m\sqrt{2\pi}} \\ &\cdot \exp\left[-\frac{1}{2}\left(\frac{i_m^{(2)}-\mu_m}{\sigma_m}\right)^2\right] di_m^{(2)} \quad (24) \\ &+ V(i_0^{(2)}+i_l^{(2)}) \int_{i_0^{(2)}+i_l^{(2)}}^{\infty} \frac{1}{\sigma_m\sqrt{2\pi}} \exp\left[-\frac{1}{2}\left(\frac{i_m^{(2)}-\mu_m}{\sigma_m}\right)^2\right] di_m^{(2)}.\end{aligned}$$

Using the notation for the cumulative normal distribution

$$N(x) = \int_{-\infty}^{x} \frac{1}{\sqrt{2\pi}} \exp\left[-\frac{1}{2}r^2\right] dr \tag{25}$$

and making the substitution

$$r = \frac{i_m^{(2)} - \mu_m}{\sigma_m} \tag{26}$$

gives

$$\begin{aligned}
I = {} & \left.\frac{dV(i_m^{(2)})}{di_m^{(2)}}\right|_{i_0^{(2)}+i_d^{(2)}} \int_{\frac{1}{\sigma_m}(i_0^{(2)}+i_d^{(2)}-\mu_m)}^{\frac{1}{\sigma_m}(i_0^{(2)}+i_\ell^{(2)}-\mu_m)} \frac{1}{\sqrt{2\pi}} \\
& \cdot \left(r\sigma_m + \mu_m - i_0^{(2)} - i_d^{(2)}\right) \exp\left[-\frac{1}{2}r^2\right] dr \\
& + V(i_0^{(2)} + i_\ell^{(2)}) \left[1 - N\left(\frac{1}{\sigma_m}(i_0^{(2)} + i_\ell^{(2)} - \mu_m)\right)\right] \\
= {} & \left.\frac{dV(i_m^{(2)})}{di_m^{(2)}}\right|_{i_0^{(2)}+i_d^{(2)}} \left\{ \frac{-\sigma_m}{\sqrt{2\pi}} \left[\exp\left[-\frac{1}{2\sigma_m^2}(i_0^{(2)} + i_\ell^{(2)} - \mu_m)^2\right]\right.\right. \\
& \left. - \exp\left[-\frac{1}{2\sigma_m^2}(i_0^{(2)} + i_d^{(2)} - \mu_m)^2\right]\right] \\
& + (\mu_m - i_0^{(2)} - i_d^{(2)}) \left[N\left(\frac{1}{\sigma_m}(i_0^{(2)} + i_\ell^{(2)} - \mu_m)\right)\right. \\
& \left.\left. - N\left(\frac{1}{\sigma_m}(i_0^{(2)} + i_d^{(2)} - \mu_m)\right)\right]\right\} \\
& + V(i_0^{(2)} + i_\ell^{(2)}) \left[1 - N\left(\frac{1}{\sigma_m}(i_0^{(2)} + i_\ell^{(2)} - \mu_m)\right)\right].
\end{aligned} \tag{27}$$

Equation (27) is evaluated for the case of the forecast from the last data point of the January 1970 to December 1978 period and for the case of the forecast from the last data point of the January 1979 to December 1984 period. Mortgages of fixed interest terms for one or five years are considered. The standard deviations of the forecast are taken from Table 3 and corrected from percentage terms by division by one hundred. Calculations are performed for two values of μ_m, the best estimate of mortgage rates at renewal. Firstly, they are made under the assumption that the best estimate is the interest rate at inception. This assumption corresponds to the practical decision by CMHC to charge a premium unrelated to recent mortgage trends. Secondly, calculations are made under the assumption that the best estimate of interest rates at renewal is given by the SARIMA forecast. The results are displayed in Table 4. In making comparisons with the CMHC premium of 1.5% of principal, it should be kept in mind that because only 75% of the principal is insured by CMHC, the premium charged corresponds to $I/RP_0 = 0.02$. The CMHC premium will also include an allowance for expenses.

Table 4. *Comparison of Computed Single Premium With CMHC Premium of 0.02*

Origin December	1978	1978	1984	1984
Rate at Origin	11.53%	11.53%	12.74%	12.74%
Term (years)	1	5	1	5
No change estimate				
Estimate at renewal	11.53%	11.53%	12.74%	12.74%
Single Premium I/RP_0	0.00015	0.00855	0.00520	0.03511
Forecast estimate				
Estimate at renewal	12.19%	12.88%	12.65%	13.41%
Single Premium I/RP_0	0.00052	0.01623	0.00501	0.03877

5. CONCLUSIONS

Table 4 indicates that the computed single premium based on the January 1979–December 1984 model is much greater than that based on the January 1970–December 1978 model. This is a consequence of the greater volatility of interest rates in the more recent period, and illustrates the difficulty of determining a fair premium if it is not clear which, if any, historical period will be the closest mimic of the future.

Since the first difference of monthly mortgage rates is autoregressive, there are trends. An upward movement in interest rates over the last few months will tend to be continued, and theoretically the premium charged should reflect this fact.

It is clear from Table 4 that the term of the mortgage has a large effect on the required premium. There are two reasons for this:

- a long term mortgage requires a subsidy for a longer period.
- interest rates tend to diverge further from their initial value with time (proportional to the square root of time for random walk).

Thus the use by CMHC of "a 1.5% for all terms" single premium is an administratively convenient approach, but one which neglects a very important factor.

ACKNOWLEDGMENTS

This research was supported by a grant from the Natural Sciences and Engineering Research Council of Canada. The author is grateful to Dr. Phelim P. Boyle for helpful comments and to Mrs. Linda Gregory for an expert typing job.

APPENDIX A

Calculation of Insurance Prices from December 1978 Origin

	$m=1$		$m=5$	
μ_m :	.1153	.1219	.1153	.1288
n	25	25	25	25
$i_0^{(2)}$	0.1153	0.1153	0.1153	0.1153
$i_\ell^{(2)}$	0.1200	0.1200	0.1200	0.1200
$i_d^{(2)}$	0.0200	0.0200	0.0200	0.0200
$a_{\overline{m}\rvert}^{(12)}(i_0^{(2)})$	0.9416	0.9416	3.8098	3.8098
$a_{\overline{n-m}\rvert}^{(12)}(i_0^{(2)})$	8.2766	8.2766	7.9357	7.9357
$a_{\overline{n-m}\rvert}^{(12)}(i_0^{(2)}+i_d^{(2)})$	7.2685	7.2685	7.0427	7.0427
$a_{\overline{n-m}\rvert}^{(12)}(i_0^{(2)}+i_\ell^{(2)})$	4.4323	4.4323	4.4017	4.4017
$a_{\overline{n}\rvert}^{(12)}(i_0^{(2)})$	8.3405	8.3405	8.3405	8.3405
$(Ia^{(12)})_{\overline{n-m}\rvert}^{(12)}(i_0^{(2)}+i_d^{(2)})$	47.9474	47.9474	43.0092	43.0092
$\frac{1}{RP_0}\frac{dV(i_m^{(2)})}{di_m^{(2)}}\Big\rvert_{i_0^{(2)}+i_d^{(2)}}$	0.7101	0.7101	1.6809	1.6809
$\frac{1}{RP_0}V(i_0^{(2)}+i_\ell^{(2)})$	0.0735	0.0735	0.1763	0.1763
σ_m	0.011708	0.011708	0.031692	0.031692
$\Phi((i_0^{(2)}+i_\ell^{(2)}-\mu_m)/\sigma_m)$	0.9999	0.9999	0.9999	0.9996
$\Phi((i_0^{(2)}+i_d^{(2)}-\mu_m)/\sigma_m)$	0.9562	0.8738	0.7361	0.5813
$\frac{1}{RP_0}I$	0.00015	0.00052	0.00855	0.01623

APPENDIX B

Calculation of Insurance Prices from December 1984 Origin

	$m = 1$		$m = 5$	
μ_m :	.1274	.1265	.1174	.1341
n	25	25	25	25
$i_0^{(2)}$	0.1274	0.1274	0.1274	0.1274
$i_\ell^{(2)}$	0.1200	0.1200	0.1200	0.1200
$i_d^{(2)}$	0.0200	0.0200	0.0200	0.0200
$a_{\overline{m}\rceil}^{(12)}(i_0^{(2)})$	0.9359	0.9359	3.7112	3.7112
$a_{\overline{n-m}\rceil}^{(12)}(i_0^{(2)})$	7.6395	7.6395	7.3739	7.3739
$a_{\overline{n-m}\rceil}^{(12)}(i_0^{(2)} + i_d^{(2)})$	6.7595	6.7595	6.5831	6.5831
$a_{\overline{n-m}\rceil}^{(12)}(i_0^{(2)} + i_\ell^{(2)})$	4.2299	4.2299	4.2057	4.2057
$a_{\overline{n}\rceil}^{(12)}(i_0^{(2)})$	7.6877	7.6877	7.6877	7.6877
$(Ia^{(12)})_{\overline{n-m}\rceil}^{(12)}(i_0^{(2)} + i_d^{(2)})$	42.2860	42.2860	38.4313	38.4313
$\frac{1}{RP_0}\frac{dV(i_m^{(2)})}{di_m^{(2)}}\Big\vert_{i_0^{(2)}+i_d^{(2)}}$	0.7085	0.7085	1.5855	1.5855
$\frac{1}{RP_0}V(i_0^{(2)} + i_\ell^{(2)})$	0.0727	0.0727	0.1648	0.1648
σ_m	0.038390	0.038390	0.085064	0.085064
$\Phi((i_0^{(2)} + i_\ell^{(2)} - \mu_m)/\sigma_m)$	0.9991	0.9992	0.9208	0.9085
$\Phi((i_0^{(2)} + i_d^{(2)} - \mu_m)/\sigma_m)$	0.6988	0.7069	0.5930	0.5621
$\frac{1}{RP_0}I$	0.00520	0.00501	0.03511	0.03877

REFERENCES

Abraham, B., and J. Ledolter (1983), *Statistical Methods for Forecasting.* New York: Wiley and Sons.

Box, G. E. P., and G. M. Jenkins (1978), *Time Series Analysis, Forecasting and Control.* San Francisco: Holden-Day.

Buser, S. A., and P. H. Hendershott (1984), "The pricing of default-free mortgages". Working paper, National Bureau of Economic Research, Cambridge, Massachusetts.

McLeod, A. I. (1978), "Manual for the TS package". Mimeograph, Department of Statistics and Actuarial Sciences, University of Waterloo.

Samuel H. Cox, Jr., and Cheng-kun Kuo [1]

UNDERWRITING TRADERS OF FINANCIAL FUTURES

1. INTRODUCTION

A futures contract is an agreement to trade an asset at a specific time in the future, at a specific price. Traditionally, the asset involved would be a commodity such as corn, gold, etc. or a financial asset such as a bond or stock. Recently, futures markets have developed for economic variables such as interest rates, freight shipping rates, and currency exchange rates. At the time the contract is made, the buyer and seller agree on the future date of the trade, called the maturity date of the futures contract, and the price of the asset to be traded called the futures price. A forward contract is identical to a futures contract, except in the way the difference between the market price of the asset on the date of the trade and the futures price is handled. Forward contracts require that the difference be made up in cash to the gaining party at the time the trade takes place. Futures contracts require that changes in the futures price be paid in cash as they occur. That is, futures contracts are distinguished from forward contracts by daily resettlement of margin variations. This is called "marking to the market". [It is interesting that until recent work by Cox *et al.* (1981) and others (French, 1983; Jarrow and Oldfield, 1981; and Richard and Sundaresan, 1981), most researchers in finance ignored this difference.]

As an illustration of the impact of marking to the market, consider International Monetary Market (IMM) million dollar ninety-day U.S. Treasury bill futures. As with all commodities, an initial margin is required. In this case, it is only $1500. The leverage is high and the corresponding risk is tremendous. For example, a 0.1% change in the futures price of $1,000,000 implies a $1000 change in the traders' accounts, i.e., a 67% change. [The IMM does stipulate minimum and maximum price changes. For any bid or ask, the minimum price change is one basis point (0.01%) and the maximum is 50 basis points (0.5%). We are ignoring such constraints on price

[1] Department of Finance, The University of Texas at Austin, Austin, Texas 78712 (both authors)

I. B. MacNeill and G. J. Umphrey (eds.), Actuarial Science, 219–229.

changes.] It is surprising that few researchers considered the cash flow strain this may imply. Morgan (1981) has indicated that daily resettlement poses no problem for hedgers, although he noted that brokers can take a loss on positions that are not covered by margin requirements. However, Powers and Vogel (1984) do warn that an investor easily could be forced to sell into an upward market at a time the market is strong and, if an immediate reversal of the price trend does not occur, substantial margin reserves may be needed to hold the position.

Thus a serious problem for a trader of futures contracts is to determine how much cash should be held to meet the requirements of marking to the market. Financial futures and financial option contracts could be used by insurance companies and pension fund managers to hedge the return on bond and stock investments, to the extent regulations allow it. Two articles by Clancy (1985) and Fen (1985) are an indication of increasing interest in these contracts among actuaries. Our model will show how an insurance company or pension plan could determine the probable effect of futures trading on its daily cash flow. This may lead to regulations which allow insurance companies to trade futures to a much greater extent than is currently the case.

The broker, through which financial futures traders deal, assures other market participants that marking to the market will be maintained. If an individual trader fails to mark to the market, the broker can take the opposite position at the end of the day to prevent further losses, but must cover the client's loss for the day. Because of this, brokers have rules for "underwriting" traders. For example, one brokerage firm requires that a customer have a net worth of at least $120,000 and an annual income of at least $30,000 to trade futures contracts. But it has no liquidity requirement nor does the credit rule reflect the nature of the trader's portfolio. This is typical. Although the broker has the problem of determining how much cash a client trader should be expected to have available for marking to the market requirements, the current practice is to use an *ad hoc* credit rating rather than to solve the problem directly. We hope that our model is a step toward the proper credit requirements set by brokerage firms so that any investor having the resources and the skill, could participate in the increasingly important financial futures markets.

To present our idea in a simple setting, we will consider a single futures contract written on a bond which pays no coupons, sometimes called a pure discount bond. It is not much more complex to generalize to coupon bonds; we will do this in a forthcoming article presenting applications of the model developed here. Futures on bonds are quite important because, according to the Futures Industry Association, the trading volume of futures (and of options on futures) on U.S. Treasury bonds is very high; 30 million futures contracts were traded in 1984. (See 'The Current State of the Futures Industry', *Futures,* October 2–5, 1985.) Eventually we will develop the model to the point that it applies to futures written on stock indices, interest rates and currencies. Our solution is subject to some rather strong assumptions:

(A1) The excess log t-period holding return for bonds follows a Brownian Bridge process.
(A2) The futures price $F(t)$ is determined by the price of the bond $P(t)$.
(A3) All investors are rational and have all the available information.
(A4) There are no taxes and no transaction costs related to futures trading.

2. THE DYNAMICS OF A PURE DISCOUNT BOND

Let $P(t)$ be the market price of the bond at time t, $0 \leq t \leq m$. At time m the bond owner redeems the certificate for K dollars. There are no coupons paid; the bonds trade at a discount. (The few times we will need to distinguish bonds of different maturities we will write $P(t,m)$ rather than $P(t)$). Viewed from time 0, $P(0)$ is known but all future market values $P(t)$ are uncertain. We model this uncertainty with a stochastic process described below. If the bond is bought at time 0 and held to time t, then the yield, in terms of a continuously compounded annual interest rate for t years, is

$$\xi(t) = (1/t)\log[P(t)/P(0)]$$

and the t-period rate is $t\xi(t)$. That is, $P(0)\exp(t\xi(t)) = P(t)$. Let $\mu = \xi(m)$ denote the yield to maturity. This is the continuously compounded rate of return, provided the bond is purchased at time 0 and held to maturity at time m. If $P(0)$ were invested in a savings account which paid, with certainty, μ per annum continuously, then the deterministic process $P(0)\exp(t\mu)$ would have the same initial and final values as $P(t)$.

When an investor buys at time 0, but does not hold to maturity, the investor's t-period yield to the time t of selling is $t\xi(t)$, which may be different from yield $t\mu$, which a bond held to maturity would earn over the same period. Define the excess return to be

$$\eta(t) = t\xi(t) - t\mu.$$

For convenience, we are now changing the time units so that we may assume $m = 1$. Note that the stochastic process $\eta(t)$, $0 \leq t \leq 1$, satisfies

$$\eta(0) = 0 \qquad \text{and} \qquad \eta(1) = 0.$$

These conditions suggest that we characterize $\eta(t)$ as a "Brownian Bridge" process, which is a standard Brownian motion process $[0,1]$, conditioned by the two constraints above. [A standard Brownian motion $Z(t)$ on $[0,m]$ has constant drift equal to zero and constant diffusion coefficient equal to one.] This, in some sense, is the simplest nontrivial diffusion process which satisfies the above requirements. Karlin and Taylor (1981, p. 269) have shown that the Brownian Bridge process $B(t)$ can be written as

$$B(t) = (1-t)Z(t/(1-t)),$$

where $Z(t)$ is a standard Brownian motion on $[0,1]$. They also show that the stochastic differential equation of $B(t)$ is

$$dB(t) = [-B(t)/(1-t)]dt + dZ(t), \qquad 0 \le t \le 1$$

subject to $B(0) = 0$. By analogy, we assume that $\eta(t) = \sigma B(t)$, where σ is a coefficient which would have to be estimated from prior observations of bond price variations. This gives us a stochastic differential equation for the excess return:

$$d\eta(t) = [-\eta(t)/(1-t)]dt + \sigma dZ(t), \qquad 0 \le t \le 1,$$

subject to $\eta(0) = 0$. Therefore,

$$P(t) = P(0)\exp(t\mu + \eta(t)).$$

Now we apply Itô's formula (Karlin and Taylor, 1981, p. 347) to obtain a stochastic differential equation for $P(t)$:

$$dP(t) = P(t)\left\{\mu - \eta(t)/(1-t) + \sigma^2/2\right\}dt + P(t)\sigma dZ(t), \quad 0 \le t \le 1.$$

Since $\mu = \eta(t)/(1-t)$ reduces to $(1-t)^{-1}\log(K/P(t))$, then we can write this as

$$dP(t) = P(t)\left\{(1-t)^{-1}\log(K/P(t)) + \sigma^2/2\right\}dt + P(t)\sigma dZ(t), \quad 0 \le t \le 1.$$

This gives us the drift and diffusion terms as functions of t and $P(t)$:

$$dP(t) = a(t, P(t))dt + b(t,p)dZ(t), \qquad 0 \le t \le 1,$$

where

$$a(t,p) = p(1-t)^{-1}\log(K/p) + p\sigma^2/2, \quad 0 \le t \le 1, \quad p > 0$$

and

$$b(t,p) = p\sigma, \quad 0 \le t \le 1, \quad p > 0.$$

The use of the Brownian Bridge process to model discount bond prices was suggested in conversations with P. L. Brockett. Ball and Torous (1983) used it for this purpose and calculated prices for options on such bonds. We will use this stochastic differential equation for $P(t)$ in the next section to derive the stochastic differential equation for a futures contract written on a pure discount bond.

3. THE FUTURES PRICE DYNAMICS

Each futures contract calls for the delivery of a bond at a specified time and at a specified price. The following example illustrates how this works out.

Example: The bond on which the futures contact is written is a \$1,000,000 pure discount bond maturing 90 days from now, time 0. (Our notation as defined above implies that we measure time in 90 day units in this example.) A trader agrees to deliver the bond 60 days from now, at time 60/90, and at a price of $F(0)$. The futures price $F(0)$, determined by the market now, is the market consensus of what price the bond will trade at 60 days from now (at $t = 60/90$). For example, if it is expected that the bond will yield 8% per annum over the last 30 days of the bond's life, then the current futures price is $F(0) = 1,000,000 \exp(-.08(30/360)) = \$993,356$. Some traders contract to buy 60 days from now at this price and others contract to sell. The contracts are arranged through a broker, who makes sure that the trades will be performed. The broker requires a small initial deposit now from each trader. Sixty days from now the market may have changed its expectation of the yield over the last 30 days of the bond's life. Suppose that the yield has increased to 8.5% per annum so the bond price 60 days from now turns out to be

$$F(60/90) = 1,000,000 \exp(-.0850(30/360)) = \$992,942.$$

Traders who wrote contracts to buy find themselves paying \$993,356 for something that is only worth \$992,942. Buyers lose \$414, while sellers gain \$414 since they receive the contract price which is \$414 over the market price. The broker requires that the trader meet future price changes as they occur, rather than at the contract's maturity. As the market's expectations about the yield over the last 30 days of the bond's life change, the traders gain and lose. In the above example, \$414 is the total gain (or loss). It would be realized over the 60 day life of the futures contract. Losses must be paid as they occur. This gives rise to the risk of ruin, the risk that a trader will fail to "mark to the market".

Consider contracts which mature at time d on a pure discount bond which matures at time $1 > d$. Contracts are made continuously over the times $0 \leq t \leq d$, all to be performed at time d, but at various prices $F(t)$. Of course, as t approaches d, $F(t)$ approaches $P(d)$. But at earlier times, we know that $F(t)$ and $P(t)$ are quite different. We assume that $F(t)$ can be written as a function of t and $P(t)$. That is, there is a continuously differentiable function $u(t,p)$ of two real variables t and p, such that $F(t) = u(t, P(t))$. Our aim now is to discover u. Then we could proceed to calculate the probability of ruin $\Psi(x)$ for a trader who contracts to buy and has working capital x:

$$\Psi(x) = \text{Prob}\left[F(t) - F(0) > x \text{ for some } t \leq d\right].$$

First we apply Itô's Formula to obtain the stochastic differential equation for $F(t, P(t))$:

$$dF(t) = h(t, P(t))dt + g(t, P(t))dZ(t),$$

where

$$h(t,p) = u_1(t,p) + a(t,p)u_2(t,p) + (b(t,p))^2 u_{22}(t,p)$$

and

$$g(t,p) = b(t,p)u_2(t,p).$$

The subscripts denote partial differentiation. To solve the stochastic differential equation for the futures price, we will follow Vasicek (1977) in his use of an arbitrage procedure proposed by Merton (1971). Consider an investor who continuously and costlessly enters simultaneously short as well as long futures contracts on identical assets, but with different expiration dates. Denote by N_S and N_L the number of futures contracts, short and long, respectively. Let F_S and F_L denote the futures price on each contract in which the investor has taken, respectively, a short and long position. The gain to the investor up to time t is

$$W(t) = -N_S(t)F_S(t) + N_L(t)F_L(t).$$

Of course, both futures prices satisfy a stochastic differential equation of the form derived above for an arbitrary futures contract. Thus we have

$$dF_S(t) = h_S(t, P(t))dt + g_S(t, P(t))dZ(t)$$

and

$$dF_L(t) = h_L(t, P(t))dt + g_L(t, P(t))dZ(t).$$

Now we consider how the value of the portfolio changes. By Itô's formula we have

$$dW(t) = -N_S dF_S + N_L dF_L - F_S dN_S + F_L dN_L - dN_S dF_S + dN_L dF_L.$$

Now we assume in a manner similar to Merton (1971) and Vasicek (1977) that the last four terms net to zero. This assumption amounts to assuming that changes in the portfolio are due solely to futures price changes. Therefore, we obtain the following:

$$dW(t) = -N_S dF_S + N_L dF_L,$$
$$dW(t) = -N_S\{h_S dt + g_S dZ\} + N_L\{h_L dt + g_L dZ\},$$

and

$$dW(t) = \{N_L h_L - N_S h_S\}dt + \{N_L g_L - N_S g_S\}dZ.$$

Now, continuing to mimic Vasicek (1977), we suppose that the portfolio is continuously adjusted so that the second term is zero. The portfolio thus constructed accumulates risklessly. We have the no riskless arbitrage result that the portfolio realizes the same return as the riskless rate r. This leads to the following:

$$N_L g_L - N_S g_S = 0, \quad \text{i.e.,} \quad N_L g_L = N_S g_S,$$

$$N_L h_L - N_S h_S = rW,$$
$$N_L h_L - N_S h_S = r(-N_S F_S + N_L F_L),$$

and

$$N_L h_L - rN_L F_L = N_S h_S - rN_S F_S.$$

Therefore,

$$\{N_L h_L - rN_L F_L\}/N_L g_L = \{N_S h_S - rN_S F_S\}/(N_S g_S)$$

or

$$\{h_L - rF_L\}/g_L = \{h_S - rF_S\}/g_S.$$

This means that regardless of the expiration date of the futures contract, for all futures contracts on a given bond in this market, the ratio $(h - rF)/g$ is a constant. Vasicek (1977) labeled this quantity the market price of risk. It is a measure of the relative risk of a futures contract on a given bond, and it depends only on the bond. We denote it by $q_P(t)$:

$$q_P(t) = (h(t, P(t)) - rF(t, P(t)))/g(t, P(t)),$$

where F is any of the futures prices of P and $dF = hdt + gdZ$. The functions h and g were obtained in terms of the parameters of P earlier:

$$h(t,p) = u_1(t,p) + a(t,p)u_2(t,p) + (b(t,p))^2 u_{22}(t,p)$$

and

$$g(t,p) = b(t,p)u_2(t,p).$$

Making these substitutions yields the following:

$$bu_2 q_P = u_1 + au_2 + b^2 u_{22} - ru.$$

It is common in writing financial mathematics, to let $F(t)$ denote both the stochastic process $u(t, P(t))$ and the function $u(t,p)$. Then, switching to another notation involving partial derivatives, we would obtain the following equivalent partial differential equation:

$$b(\partial F/\partial P)q_P = \partial F/\partial t + a(\partial F/\partial P) + b^2(\partial^2 F/\partial P^2) - rF$$

or

$$\partial F/\partial t + \{a - bq_P\}(\partial F/\partial P) + b^2(\partial^2 F/\partial P^2) - rF = 0.$$

We prefer to maintain distinction between $F(t)$ (which exists independently of anyone's assumptions) and the conjectured function $u(t,p)$. We do adopt the alternate partial derivative notation. Therefore we seek a function $u(t,p)$ which satisfies the partial differential equation

$$\partial u/\partial t = \{bq_P - a\}(\partial u/\partial p) - b^2(\partial^2 u/\partial p^2) + ru$$

subject to the condition that $u(d,p) = p$ for all p, where d is the expiration date of the futures contract. This boundary condition was discussed earlier. Using a method similar to Vasicek's (1977), it can be shown that the solution can be obtained as

$$u(t,p) = E\left[P(d)\exp(-\Lambda(t)) \mid P(t) = p\right],$$

where

$$\Lambda(t) = \int_t^d r(\tau)d\tau + \frac{1}{2}\int_t^d q^2(\tau)d\tau + \int_t^d q(\tau)dZ(\tau).$$

An approximation could be obtained by simulation or other numerical methods. In either case, we can approximate the distribution of $F(t) = u(t, P(t))$. From this we can estimate the effect on futures price variation. Suppose a trader starts with initial working capital x. At time t, the trader's working capital will be

$$W(t) = x + F(t) - F(0)$$

and the probability of ruin before expiration of the contract is

$$\Psi(x) = \text{Prob}\{W(t) < 0 \text{ for some } t \leq d\}.$$

Actuaries will recognize the determination of $\Psi(x)$ as a ruin theory problem. There are several well known methods in the actuarial literature for approximating $\Psi(x)$ in terms of parameters of W and x . A solution to the problem we introduced could be obtained by one of these methods. That is, this model can be used to determine the value of x for which $\Psi(x)$ is acceptably small. Conversely, it could be used to determine the probability $\Psi(x)$ of failure to mark to the market given the amount x of initial working capital. We expect to present numerical results of this method in the near future. We have learned recently that Kolb *et al.* (1985) have presented a model in which the futures price, $W(t)$ in our notation, is assumed to be a Brownian motion with zero drift parameter and diffusion coefficient σ. In this case, the probability of ruin is simply $\Psi(x) = 2(1 - \Phi(x/\sigma\sqrt{d}))$, where Φ denotes the distribution function of a standard normal random variable.

4. THE OPTIMAL LEVEL OF WORKING CAPITAL

We note that for traders who anticipate increasing the amount of working capital if it is necessary to maintain a contract, $\Phi(x)$ is not of direct concern. The problem is then really one of cash management. We are thinking that the trader would initially dedicate x to meet margin requirements, and every time $W(t)$ sank to 0 it would immediately be reestablished at the initial level x. A cost of reestablishing the fund would be incurred, and there would be a loss of foregone interest since the cash held for margin requirements would not earn as much as less liquid investments. If x were too low, then there would be excessive transaction costs. If it were too high, the foregone interest would be too great. Thus there must be some optimal level x of cash a trader should maintain. This is similar to the setting used by Frenkel and Jovanovic (1980) to determine the optimal working capital a firm should dedicate for its business transactions. However, in their article, the stochastic process analogous to our $W(t)$ is a Brownian motion with negative drift and they supposed that the time period was indefinite. In our case $W(t)$ need not be a Brownian motion, and the time period involved, $[0, d]$, is finite. Moreover, we think it is reasonable to consider that the trader may withdraw cash from the working capital fund if it reaches such a high level y that it is safe to bring it down to the level x. Of course this is costly also. For simplicity, we suppose that both transactions costs are the same, i.e., k dollars per transaction.

In futures trading, the cash flows needed for daily resettlements may be transferred from a checking account which earns a constant riskless rate of interest r. The relevant interest foregone, then, is the difference between r and the most likely rate earned if the capital is put to better use. A natural candidate for this probably earned better rate is the rate which an investment on bonds would earn over the term of the futures contract. This is $\mu(d) = [\log(P(d,d)/P(0,d))]/d$ and interest is lost at the rate $\mu(d) - r$.

Now that we have identified the costs of maintaining a futures contract, we should calculate the present value of these costs at time 0, and choose x and y to minimize it. The numerical solution of this optimization is technically possible, but quite tedious. This is because our process is not necessarily Brownian, and we have a finite interval $[0, d]$ for managing the fund. The problem is somewhat easier if operating the fund indefinitely is contemplated. In fact, many traders of futures on bonds never take or make delivery of the bonds, but rather take an opposite position, closing out the current obligation, and entering into a new contract with a later expiration date. The analogous cash management problem (i.e., with indefinite horizon) was worked out by Lee and Cox (1982) for the case where $W(t)$ is a Brownian motion, generalizing the results of Frenkel and Jovanovic (Following Frenkel and Jovanovic, the drift and diffusion coefficients are denoted respectively by $-\mu$ and σ, but because we have two barriers, it is not necessary to assume $\mu > 0$ in order to obtain a solution.) The solution in this special case can be viewed as an approximation to the general solution. The present value of costs of trading futures indefinitely in this case is

$$G(x,y) = \{x + k\alpha\}/(1-\alpha) - \mu/r,$$

where

k is the cost per transaction,

$\alpha = E[\exp(-rT)]$,

$T = \min\{T(0), T(y)\}$,

and

$$T(z) = \inf\{t \geq 0 \mid 0 < W(s) < y \text{ for all } s < t\}, (z = 0, y).$$

5. SUMMARY

We have presented a framework for the analysis of cash requirements for traders of financial futures. The basic notion is that the futures price should be a function of the current market price of the good to be delivered. We selected discount bonds as the good, and used a Brownian Bridge process to model bond prices. Formulas were derived for relating initial working capital to the likelihood that it will fail to be adequate. We also addressed the problem of finding an optimal procedure for trading futures contracts indefinitely. Numerical values will be presented in the near future.

REFERENCES

Ball, C. A., and W. N. Torous (1983), "Bond price dynamics and options". *Journal of Financial and Quantitative Analysis* **18**, 517–531.

Clancy, R. P. (1985), "Options on bonds and applications to product pricing". *Transactions of the Society of Actuaries* **37**, 97–130.

Cox, J. C., J. E. Ingersoll, Jr., and S. A. Ross (1981), "The relation between forward prices and futures prices". *Journal of Financial Economics* **9**, 321–346.

Fen, A. M. (1985), "Interest rate futures: an alternative to traditional immunization in the financial management of guaranteed investment contracts". *Transactions of the Society of Actuaries* **37**, 153–184.

French, K. P. (1983), "A comparison of futures and forward prices". *Journal of Financial Economics* **12**, 311–342.

Frenkel, J. A., and B. Jovanovic (1980), "On transactions and precautionary demand for money". *Quarterly Journal of Economics* **95**, 25–43.

Jarrow, R. A., and G. S. Oldfield (1981), "Forward contracts and futures contracts". *Journal of Financial Economics* **9**, 373–382.

Karlin, S., and H. M. Taylor (1981), *A Second Course in Stochastic Processes*. New York: Academic Press.

Kolb, R. W., G. D. Gay, and W. C. Hunter (1985), "Liquidity requirements for financial futures investments". *Financial Analysts Journal*, May–June, 60–68.

Lee, W. Y., and S. H. Cox, Jr. (1982), "The impact of upper and lower control limits on the transactions and precautionary demand for money". Department of Finance Working Paper, University of Texas at Austin.

Merton, R. (1971), "Optimal consumption and portfolio rules in a continuous time model". *Journal of Economic Theory* **3**, 373–413.

Morgan, G. E. (1981), "Forward and futures prices of treasury bills". *Journal of Banking and Finance* **5**, 483–496.

Powers, M., and D. Vogel (1984), *Inside the Financial Futures Markets*. New York: Wiley and Sons.

Richard, S. F., and M. Sundaresan (1981), "A continuous time equilibrium model of commodity prices in multiperiod economy". *Journal of Financial Economics* **9**, 347–372.

Vasicek, O. A. (1977), "An equilibrium characterization of the term structure". *Journal of Financial Economics* **5**, 177–188.

D. S. Rudd, F.S.A., F.C.I.A. [1]

MERGERS OF LIFE COMPANIES AND THE BLURRING OF BOUNDARIES AMONG FINANCIAL INSTITUTIONS—EFFECTS ON THE ACTUARIAL PROFESSION

Recent events relevant to the topic of this discussion are the takeover of Monarch Life by North American Life and the takeover of Dominion Life by Manulife, in both cases for subsequent merger. Also, there has been considerable discussion concerning the powers of financial institutions. These events and discussion present a possible picture of both contraction and expansion of the need for actuaries.

The Life Insurance business has changed drastically over the years. Attempts by government to keep it separate from other financial fields and prevent concentration have failed. The steadily decreasing mortality risk, the rise in volatility of interest rates, and the attraction of managing large pools of capital, have changed the nature of the life business and its ownership. Profits from the spread in interest rates, from investment management, and from managing to avoid taxation, have become much more significant than mortality profit which is under ever-increasing and more sophisticated competition for the pure insurance dollar, individual or group. Financial conglomerates, of which once there were only one or two, are now supposedly the ideal. In the trust industry, self-dealing and recklessness have brought the demise of trust companies. Life companies are not immune to this style of management; for example Paramount Life, an Alberta Company went down the tube loaded with poor investments.

We now consider mergers. Unlike other financial institutions, Life Insurance companies for decades were not permitted to grow by acquisition/merger as were trust companies and banks. Power Corporation was not permitted to merge Imperial Life into Great West Life despite almost complete ownership of both companies. Only a couple of years ago, the

[1] Department of Statistical and Actuarial Sciences, The University of Western Ontario, London, Ontario N6A 5B9; also President of D. S. Rudd Associates Ltd., Consulting Actuaries, Suite 412, 200 Queens Avenue, London, Ontario N6B 1V5

I. B. MacNeill and G. J. Umphrey (eds.), Actuarial Science, 231–234.

Laurentian Group was not permitted to merge Northern Life into Imperial, which they both then owned. A federally registered Life Insurance company was not permitted to acquire another life company in Canada—unless to rescue it from probable insolvency.

Two years ago North American Life purchased the stock company Monarch Life of Winnipeg, a form of indirect mutualization, for merger. Last year Lincoln National's holding company put Dominion Life up for sale. It was purchased by Manulife which received permission to merge Dominion Life into its operations; the purchase price in my opinion, could not be justified except on some merger basis with large potential savings in overhead. The rules of the game have been changed. The value of a life company has been increased and presumably we shall see more small and medium sized companies bought up by larger more efficient companies.

Obviously this will tend to reduce the employment possibilities for actuaries. This could be offset to some degree by the tendency of larger companies to use actuaries in non-actuarial roles of the company; however, this tendency may not continue.

The restructuring of financial institutions and the use of holding companies to own several life insurance companies has been and still is an alternative to merger. However, it would seem to be an uneconomical way of operation unless there are special reasons, such as for regional companies or for specialty companies. Examples are a Quebec company for a French-speaking nationalistic clientele, and a disability income company because of the very different pricing and claim problems of that business.

FINANCIAL CONGLOMERATES

Power Corporation's long time holding through Investors' Group of Great West Life and Montreal Trust is probably the best known conglomerate. More recent is the development of the Laurentian Group with holdings of mutual funds, life insurance, casualty insurance, trust and banking operations. Here in London, we have seen London Life become part of Trilon which was formed by Brascan. Trilon now has a substantial interest in Royal Trust and A. E. Lepage (real estate), and recently purchased the Canadian casualty insurance operations of Firemen's Fund, now renamed Wellington Insurance. Earlier the Prenor Group, which owned a trust and casualty company, purchased Northern Life. It got into financial difficulties through its casualty company, and sold off its two insurance operations to Laurentian. The combination, in other words, sometimes doesn't work.

Currently, "synergy" through full financial services is back in vogue after a flurry in the 1960's. The movement has been greater in the U. S. For example, Prudential of America not only formed a casualty company to sell fire and auto (mainly to help support its agents), but subsequently moved into mutual funds, brokerage operations, and like most larger companies is into the investment management field as a spin-off from managing corporate pension funds.

The regulatory climate in Canada has not permitted a federally registered life insurance company to enter directly into these lines except those that were "ancillary" to its chartered functions. For example, permission to form a data processing or an investment management subsidiary could be obtained. However, the holding company route enables the same group of shareholders to enter into wider fields, albeit without easy access to the capital and surplus of the life company itself! Quebec-based companies had greater flexibility and were recently freed to do much more—placing mutuals on a basis similar to stock companies who had available the "upstairs" holding company route used with London Life.

The Honourable Barbara McDougall, Minister of State (Finance) released in 1985 a discussion paper on the "four pillars" of the financial sector in Canada. The four pillars are the chartered banks, the trust companies, the life insurance companies and the investment dealers. The latter group does not manage and collect large pools of assets, but moves the money around. Our dual jurisdiction of provincial and federal governments makes regulation much more complex than in the U. S. or the U. K. The relatively large size of the chartered banks and their extensive branch networks are feared and envied by the other three pillars, who generally oppose any expansion of powers for the banks, while calling for more powers for themselves.

The Discussion Paper sets out the three main rationales for regulation:

–solvency, stimulated by recent failures,
–self-dealing and conflicts of interest,
–competition and concentration.

The solution proposed by the Discussion Paper is to adopt and enlarge on what is already happening with organizations such as Trilon; that is, the adoption of a policy of allowing the existence of financial holding companies, including wholly owned new "Schedule C" banks, with greatly increased rules against self dealing. The "Schedule C" bank would indirectly allow the other financial institutions into the commercial lending field, and the greatly enlarged self dealing rules would attempt to prevent disasters such as Seaway and Crown Trust, Paramount Life, and to limit the damage done to Northern Life as it changed hands four times with new owners making use of its assets. Mutual Life insurance companies would be allowed to form "downstairs" holding companies to place them on an equal footing. Each type of financial institution would continue to have its own appropriate regulatory approach to solvency.

Now where does all that lead to with respect to actuaries? Their traditional roles in the life insurance companies will no doubt remain. However, a recent editorial in the "Actuary" magazine of the Society of Actuaries indicated that only 8% of the life company CEO's in the U. S. were actuaries. 20% of the largest sixty-one companies had an actuary as CEO, but this includes almost all the major mutuals, which are management controlled.

At one time years ago, the actuaries in the life company probably represented 80 to 90% of the university graduates in the company. Business was simple, attitudes were conservative, and solvency was most important after the troubles of the 1930's and the low interest rates of the 40's, when rates were only in the range of 3% to 6%. The explosion of group benefits, and more recently development of non-traditional products and heightened competition has certainly made for more actuarial staff. However, the increasing importance of marketing and general financial and business acumen is leading to fewer actuaries in the top management positions. The proposed solution of permitting the existence of conglomerates of financial institutions, rather than expansion of powers for the life company, may tend to hive actuaries more into traditional insurance work in the life company.

On the other side of the coin is a greater tendency for actuaries to play a role in financial matters in general; this tendency leans towards the British experience where actuaries historically have not only been responsible for both sides of the balance sheet, but have played roles in the financial community, for example, by being on the staff of brokerage and investment management houses.

Inside the life company the actuary is much more involved now than in the past with the matching of assets and liabilities, the calculation of investment risk, and such. Possibly this talent in the life company will cross between affiliates to the other financial institutions in the stable. Casualty insurance affiliates are another example where life company actuaries, with the necessary additional qualifications, may assume non-traditional roles since there are few casualty actuaries in Canada.

On balance however, I think there may be a decrease in the scope for actuaries, at least for moving into higher management positions. Part of this feeling arises because of the higher mathematical skills and abilities which seem to be demanded from today's actuarial students. Today's actuaries have statistical and mathematical skills far beyond those demanded in my day. They are tackling problems in risk theory beyond anything we considered. Their technical skills are far greater than we had.

However, will the new actuaries have the broader general education and skills for the new environment? Or will they tend to become more the backroom boffins and technicians of the life company?

K. W. Stewart, F.S.A., F.C.I.A. [1]

COMMENTARY ON "MERGERS OF LIFE COMPANIES AND BLURRING OF BOUNDARIES BETWEEN FINANCIAL INSTITUTIONS—EFFECTS ON THE ACTUARIAL PROFESSION"

EVOLUTION OF FINANCIAL SERVICES

One of the laudable postulates of the McDougall Green Paper is that it explicitly recognizes the reality of change—not just that change will occur in the future (as indeed it will), but that change has already been upon us, and continues to accelerate.

When people talk about the shootout between financial institutions, they often express the fear that large companies will become even larger, while smaller companies are either relegated to the backwaters or are swallowed up. In the minds of the fearful, this means reduced competition and limited customer choice.

We will indeed see a significant increase in merger and acquisition activity in financial services, with both stock and mutual companies playing a major role. However, this will be accompanied by an increase in competition, expansion of alternative distribution channels, and additional choices for consumers.

Experience of the last decade, and comparisons with the U. S., noting important differences between our countries, suggest the following expectations:

1. With the increasing cost of the technology investment needed to do business in today's financial marketplace, and increasing competition across institutional and national boundaries, many smaller and mid-sized companies will combine forces by merger or acquisition to build or join larger companies. In so doing, they will be able to compete

[1] Investment Department, London Life Insurance Company, 255 Dufferin Avenue, London, Ontario N6A 4K1

I. B. MacNeill and G. J. Umphrey (eds.), Actuarial Science, 235–238.

more effectively, share distribution systems and customer franchises, and control what would otherwise be runaway operating costs.

2. While much is made of the superstore or financial supermarket, evidence from marketing studies shows that consumer markets are becoming fragmented. Instead of a monolith, THE MARKET, what we really have is a number of sub-markets, each requiring recognition of its special requirements.

   This fact suggests a continuing role for specialty companies serving sub-markets, however defined, or limiting their activity to products in which they have a demonstrated ability.

   Financial conglomerates (or supermarkets) are here to stay, but so are the specialty companies, and new companies will be formed.

3. The real losers will be the "middle class" of companies, those which do a number of things, but nothing with particular distinction. They are the endangered species among financial service companies. Some will adapt and survive as recognizable entities. Many of these companies will disappear by merger or acquisition, proving as they vanish that they have outlived their economic utility.

4. The future belongs to those who innovate and adapt, build on their strengths, understand and manage around their weaknesses.

LESSONS FOR THE ACTUARIAL PROFESSION

All of this suggests some lessons for the actuarial profession. Lines between professional practice are blurring and new players are coming on the scene. There is competition from within and without.

MBA's with heavy finance and computer skills are being turned out of business schools in increasing numbers. More and more we see people with graduate degrees in mathematics fast-tracking their way into the actuarial profession. Algebraic topologists are becoming actuaries and investment bankers. At the same time, chartered accountants are becoming more of management consultants, and less of bean-counters.

An actuarial designation and a history of faithful service in your company is no longer a pass-key to the executive suite. Nor is it a guarantee that you won't be one of the casualties of corporate reorganization. Some of the functions previously performed by actuaries can be carried out quite effectively by others, or programmed on a micro-chip.

Many actuaries have had their noses buried in the liability side of the insurance balance sheet so long that they haven't noticed the world sprinting by. To some degree the actuarial profession as a whole needs to play catch-up. But actuaries also have an opportunity to forge new roles in financial services and general business.

EVIDENCE OF CHANGE

More and more North American actuaries are becoming deeply involved in asset/liability management, that intricate and largely undiscovered relationship between the two sides of the balance sheet. I recently served on a Task Force of the Education and Examination Committee of the Society of Actuaries updating Part 8 of the Fellowship examinations, formerly economics and investments. Asset/liability management is now the unifying theme of the Part 8 examination, although not all of our recommendations have yet been implemented.

Jim Tilley, the youngest-ever member of the Board of Governors of the Society of Actuaries, and a Harvard physics Ph.D., is a Vice-President with the investment banker, Morgan Stanley in New York. Jim's team, which includes financial economists, develops and markets sophisticated risk management techniques, many of them based on the creation of synthetic or hybrid securities—the financial equivalent of genetic engineering.

Among Canadian actuaries, we find the Vice-President, Corporate Development, of Mutual Life, and of Crownx. I am Investment Planning Officer at London Life.

Page proofs for the *Transactions of the Society of Actuaries*, circulated in April, 1985, contain two papers which are indicative of an emerging breed of actuaries. The first paper developed a theoretical approach to the pricing of bond options, and used option theory to examine and evaluate risks in product design and pricing. The second paper used interest rate futures to manage investment portfolios supporting Guaranteed Investment Contracts.

The evidence is there that some actuaries are stepping well beyond the traditional bounds of actuarial practice—into investments, marketing, sales, and even federal politics.

There is no doubt that the younger generation of actuaries have a better mathematical and technical education than their predecessors, as Bill Rudd has noted. More than ever before, they come to us at or near Associate level in the examinations. They step through the Fellowship examinations so quickly that the Canadian Institute of Actuaries now imposes an experience requirement to ensure some level of practical experience outside the classroom.

OUTLOOK FOR THE FUTURE

Rapid progress through the examinations, and competition for resources among Department managers have led to changes in the way we employ our students and young actuaries. We increasingly rely on para-actuarial personnel, accountants and business graduates, or just a better educated clerical staff to do things that actuaries and students used to do.

I agree with Bill Rudd that traditional actuarial jobs will decrease in number. Some will be phased out as the present incumbents retire by pension or out-placement. Others will simply be lost to the competition as managers realize that they can hire or promote someone else to do the job.

Yet just as others can successfully enter traditional actuarial territory, so we have seen that actuaries can also reach out into new fields.

My forecasts for the future are as follows:

1. The demand for more specialized actuarial work will increase in line with the increase in the level of competition and the growing complexity of business processes.
2. This demand will be balanced by a decline in some of the traditional actuarial roles, and some displacement by other professions.
3. The blurring of boundaries between financial institutions will have a counterpart in the blurring of distinctions between the professions.

   Non-traditional actuaries will combine a better level of technical training with expertise in areas such as economics, investments, marketing and finance. As financial holding companies rationalize their structures and staffing, this new breed of actuaries will compete successfully for roles in banks, trust companies and conglomerates. Actuaries of the future will be management consultants, investment bankers, marketers, and entrepreneurs.

   Some of them will be executives or CEO's beside other people whose "initial training" was as an accountant, lawyer or banker.

As for companies, so for individuals. The future belongs to those who innovate and adapt, build on their strengths, and learn to accommodate or offset their shortcomings.

G. R. Dinney [1]

THE SEARCH FOR NEW FORMS OF LIFE

1. INTRODUCTION

There seems to be a congenital tendency for life insurance professionals, especially those who are actuaries, to try to devise new life forms. Of course, I am referring to the forms that life insurance products take. Whether there are new life forms—somewhere out there—is the subject of this paper, and this discussion is conducted in the context of a product that I devised in 1962 called Universal Life.

Universal Life is not well understood and part of the mystery about it may well be due to a failure in my communication, a failure which I will try to remedy in this article.

It can be noted that the search for totally new life forms is meaningless. This is because Universal Life was developed in 1962 *as a generic plan,* which means that it subsumes all other life insurance products. What I would like to do in this paper is to give you the total meaning of true Universal Life and then speculate on what it implies for us as professionals and indeed for the life insurance industry at large.

The search for new forms of life insurance betrays a tendency toward what might be called product stereotyping, and leads to more complication than necessary. On the other hand, simplicity is the very heart of Universal Life.

Today, as Universal Life has become the dominant form of Individual life insurance products, one hears from companies and from actuaries who say they are developing new life forms. The current buzzword is the so-called variable universal life. Other actuaries talk about "interest sensitive products". Most of these claims for "newness" are based on product stereotypes.

What I said in 1962 is that a simple and flexible product is the singular key to simplifying everything that relates to life insurance. Today, life insurance companies are using Universal Life as just another portfolio product and

---

[1] Great-West Life Assurance Company, 100 Osborne St. N., Winnipeg, Manitoba R3C 3A5

I. B. MacNeill and G. J. Umphrey (eds.), Actuarial Science, 239–250.

are missing opportunities to achieve significant productivity improvements and cost efficiencies, not only in the manufacturing part of life insurance but more significantly in the distribution part as well.

Until recent times, the structure and the psychology of life insurance has been dominated by the permanent, fixed-currency product line. It is therefore not surprising that, as in the case of investment and marketing activities, the administration of life insurance professionals has inculcated traditional value systems so that the management and administrative staff of life insurance companies tend also to be rigid, conventional and anachronistic.

In its simplest form, Universal Life is a fund, into which premiums are deposited on a regular or irregular basis and against which term insurance costs and policy expense are charged, leaving a residuum that can be used to purchase savings or paid-up annuities or paid-up life insurance, under contractual terms. Because a deposit fund or a suspense account has many potential uses, it can be used to fund miscellaneous financial products and services, even including life insurance stereotypes such as Whole Life and 20-Year Endowment. It permits virtually limitless variations, with or without interest, mortality and expense guarantees.

In schematic form Universal Life works in this way:

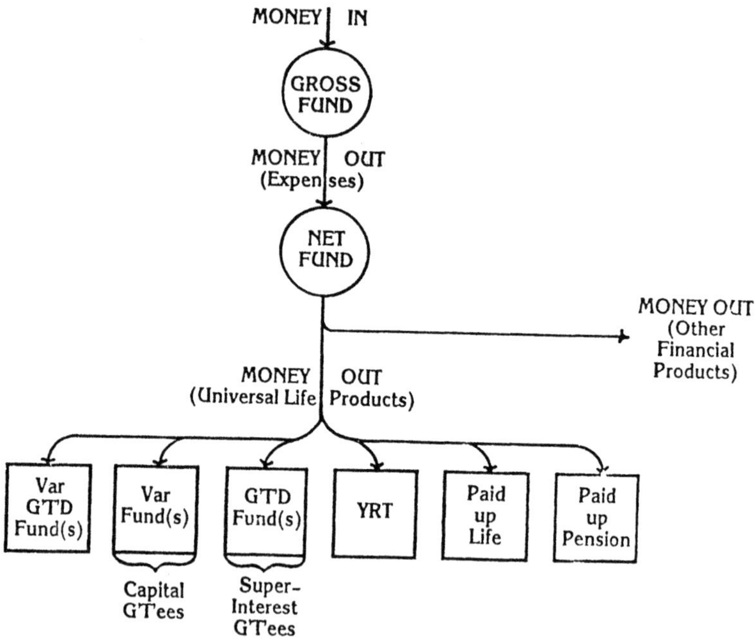

2. COMPONENTS OF UNIVERSAL LIFE

What is Universal Life and how does it work? Universal Life, being universal or unchanging, is the same today as when I devised it in 1962. It is made up of four basic elements: Term Insurance, Savings, Life Conversion, and Annuity Conversion. From the beginning I specified that the savings element had a number of options: Accumulation at a fixed rate, Accumulation at a variable rate, Accumulation at a variable/fixed rate, Plus capital guarantee, and, Plus super-interest guarantee.

Normally, I speak of true Universal Life as being a generic product rather than a product stereotype. However, under cross-examination, I might admit that all of the elements of The Universal Life Plan can be regarded as a special kind of stereotype. For example, a savings fund—such as a mortgage fund—could be regarded as a product stereotype. But it is what might be termed a *basic* stereotype and it represents one of an endless variety of such stereotypes. The fact that it is a savings stereotype does not compromise the universal concept. It is the unlimited variety of savings stereotypes that defines universality in combination with the flexibility to change easily from one to another. YRT is a product stereotype but it is the *lowest common denominator* stereotype for life insurance.

You might respond that the lowest common denominator for life insurance is not YRT but perhaps monthly renewable term (MRT) or even hourly renewable term (HRT), or instantaneously renewable term. However, that would be niggling and it does not change the idea that there is something that we might call the lowest common denominator for life insurance and that, in practical terms, it is YRT. In the case of the settlement option (for the paid up pension product), and the conversion factor (for the paid up life insurance product) we are dealing with what might be called the *ultimate* stereotype in the end game of life insurance. Therefore, we would conclude that The Universal Life Plan is modular and infinitely variable even though its component parts can be regarded as a kind of stereotype.

Many in academic circles are driven by the exhortation to "publish or perish". Before going on to describe the broader implications of Universal Life, I have three suggestions for academic enquiry and research that could make a singular contribution to the understanding of modular products.

The first area of research is one that could change our proclivity to stereotype. Why do actuaries tend to stereotype? The answer is that actuaries of my advanced years cut their teeth on Spurgeon. Younger actuaries were weaned on Jordan. I understand there is now a sequel text to Spurgeon and Jordan, titled Actuarial Mathematics, which has been co-authored by a consortium of eminent actuaries. Both Spurgeon and Jordan placed great emphasis on stereotypes, as for example in illustrating net premium calculations. It would surprise me if the new text did not do the same. I would be even more surprised if there was much reference to variable/fixed interest rates or interest-sensitive products. Those who must publish or perish might be encouraged to prepare an actuarial note, which will explain the relationship between the conventional or stereotyped calculation and modular calculation such as underlies Universal Life.

The second actuarial note would demonstrate how super-interest guarantees could be calculated in practice. The idea of a super-interest guarantee is that a policyholder may want a guaranteed interest rate that is higher than the normal interest rate on a nonpar product. This super-interest guarantee can be provided—for a charge, which would come out of the policyholder's account as a risk charge. It is easy to show how the procedure would work at the extreme. If a policyholder wants an extra 1%, the company could merely charge an additional 1% to his transaction acount. This sounds like a gold brick approach. But, with actuarial refinement, it should lead to charges of less than 1%, based on the quality of the underlying asset portfolio and the beta factor. The idea of a super-interest guarantee may not seem to be very important in practice but it could give an insurance company a competitive advantage. Moreover, the idea is highly important in extending the flexibility of Universal Life and thereby in destroying conventional myths about life insurance product and life insurance premiums. Universal Life is an attempt to meet customer needs in creative ways and that is where I come from in designing the product.

The third area of potential research lies in gross premium calculations. As products become disassembled or unbundled, it becomes vitally important to identify those cost fragments that make up the loading in gross premiums. The reason is that, under market conditions, each of these cost fragments—administration cost, tax, risk management cost, claims cost, investment cost—must bear scrutiny when compared against the comparable cost fragments for the same services provided by other management or financial intermediaries. Disassembly of product and disassembly of cost have been characteristic of the Group business for many years. In the future it will become equally characteristic of what we call Individual business.

In the search for new forms of life, most product actuaries follow a "bells and whistles" approach. This produces small, technical changes which are cosmetic rather than fundamental and which are aimed at product differentiation. This approach is normally classified under the buzzword heading VALUE ADDED. Value Added is a euphemism to describe an attempt to prolong product life under circumstances where the product is in a mature or terminal state. So, Value Added is an attempt to bring something back to life rather than an attempt to create new forms of life.

We actuaries have taken a simple idea, life insurance, and at some considerable cost we have complicated it by creating an enormous proliferation of stereotypes—like Whole Life, Twenty-Year Term, Endowment at Age 65, and so on. The cost and the complication represent Value Subtracted. Then, we put these complicated stereotypes into the hands of our sales force, which "decomplicates" the product for the benefit of the customer, but again at considerable cost. In no way am I diminishing the role of sales personnel. To the contrary, I repeat what I have been saying since 1962—that a simplified product will produce enormous simplifications in recruitment, training, retention of sales personnel, with real Value Added to the companies, to the salesmen, and above all, to our customers.

If a conventional life insurance product does not provide Value Added, where does Value Added come from in today's environment? I have two suggestions:

Firstly, through Universal Life it is possible to design products that precisely fit the needs of the customer. Moreover, it is possible to extend product design enormously through a combination of modules, including such specialized techniques as super-interest guarantees and capital guarantees. From the insurance company's standpoint it would seem to be the ideal way to create product differentiation. Moreover, modular product has the flexibility to adapt continuously to changing financial needs. Perhaps the topmost level of Value Added is in the area of auto-design or customer design of financial products. I presented an outline of this to a meeting of the Society of Actuaries held a few years ago at the University of Manitoba. My presentation was titled, "Life Insurance as a Game" (Dinney, 1982a). Quite apart from any other value that customer design will provide, it permits unprecedented brand identification or brand preference. The reason is that the customer designs the product and sells it to himself or herself, so that the desired attributes of customer loyalty and brand preference are present at time of sale and throughout the life of the policy.

There is a fundamental reason why the notion of Value Added, by creating brand preference, is inappropriate for conventional forms of life insurance. We can make this clear by comparing the marketing of life insurance and the marketing of commodities. In the merchandising of commodities, it is common practice for manufacturers to increase their price flexibility by cultivating brand preference among buyers. The contrary practice is to sell generic or unbranded products strictly on the basis of price competition.

How do life insurance products fit the generic mold? To begin, life insurance companies have tried to achieve brand preference or Value Added (and hence to avoid the price inflexibility of generic consumer products) by creating products of infinite variety. Despite the specificity and range of traditional life insurance products (e.g., Ordinary Life, Term–20, Endowment at 65 and so on) the life insurance product is easy and inexpensive to copy. As a result the most popular stereotypes are in widespread use. Because most life insurance companies offer identical products, price competition (typical of generic products) is intense. Therefore, the attempt to gain price flexibility through product specialty has merely resulted in product sameness, which contradicts the objective of creating brand preference.

In the case of life insurance product-stereotypes, the cost of inventory is low whereas the cost of product maintenance and service is high—and the exact opposite is true in the case of generic consumer products. Thus, the traditional life insurance product is a hybrid. It is designed for specificity, and hence price flexibility, but it has the worst features of generic consumer goods products—namely sameness and price inflexibility but with the added disadvantage that the traditional life insurance product is costly whereas the generic consumer goods product is cheap. In other words, the stereotyped life insurance product has all the disadvantages of the generic consumer goods product and none of its advantages. On the other hand, the modular life insurance products (like the Universal Life plan) is generic in the true sense of being a general-purpose, and not a specific-purpose, product. Brand preference for the true generic life insurance product results because of product utility and product flexibility. As previously explained, the product can be designed by the customer, thereby avoiding the sameness associated with both the life insurance stereotypes and the consumer goods generics. The consumer can also modify the product after the original point of sale. Moreover, the initial cost and the maintenance cost are low compared to the stereotyped life insurance product. The insurance company has considerable price flexibility in the case of the modular product by reason of the fact that the product modules are individually priced. Because these modules can be permuted so as to produce a multiplicity of unique life insurance products, price comparisons cannot be made easily.

Secondly, life insurance is different in one important respect from most commodity products. People do not buy life insurance because it tastes good or smells good, because it has nice texture, because it is appealing to the eye or because it sounds appealing. Life insurance is what might be termed an ethereal product, and its insubstantiality seems to be increasing inasmuch as the business has a high turnover rate and the policy benefits are running at around 60% or less of premiums.

3. PSYCHIC NEEDS AND LIFE INSURANCE

I am intrigued by the observation of the Canadian communications expert Gordon Thompson of Bell Northern Research who has said that in a post-consumption society, because physical wants are satisfied or satiated, people will turn to the acquisition and consumption of psychic goods, specifically knowledge, as a source of wealth and satisfaction. On the surface, this tendency augurs well for life insurance. More than any other financial product, life insurance provides psychic value to its customers. This psychic value is best represented by the "protection" element of the product as distinguished from its "savings" element, whose value is more tangible.

Recognition of the psychic value of life insurance in relation to the psychic needs of the public could open up, as yet, unexplored opportunities for merchandising life insurance. In this context, more emphasis would have to be given to the communication of the product idea rather than to its physical distribution to the market. The underlying concept of life insurance is ideational. It deals in abstractions. Even the most tangible benefit of life insurance—money—is itself an abstraction. An ethereal product should lend itself to distribution through information or communications technology rather than through conventional distribution.

What are some of these psychic needs that life insurance products could address? A clue to the answer is found in the book The Study of The Future, edited by Edward Cornish. In this book, reference is made to what is termed Scarcities of Abundance. The suggestion is made that pecuniary wealth and a consumption society, which are characteristic of our North American living, actually cause deprivation of certain non-tangible or psychic goods. These Scarcities of Abundance, or psychic goods, to which life insurance might address itself include the following list taken from The Study of the Future:

> *Time.* Due to complicated life-styles (pleasure boats, tourism, dinners in restaurants, etc.), people find they lack time to do everything they want to do; hence, time becomes a major constraint on their activities. By contrast, people in earlier periods of history had more time because they lacked the money that allows one to have a complex life style.
>
> *Recognition.* The amount of recognition accorded the average (median) person in western society has declined as population has grown and the mass media have become all-pervasive....the average person goes unrecognized.
>
> Western society distributes recognition even less evenly than it distributes wealth. A famous television personality may now be known to 200 million persons whereas the average person may be known to only a few hundred. The maldistribution of admiration and attention may easily be 10 or 100 times greater than the maldistribution of wealth.
>
> *Wisdom.* Growing complexity and the glut of information, most of which is irrelevant, have made it extremely difficult to obtain the information required for wise decisions.
>
> *Influence.* The growing scale of organizations has made it increasingly

difficult for individuals to have a recognizable impact on our various social systems, so the individual finds institutions "unresponsive".

Intimacy (Love, Friendship). The attenuation of community ties, the substitution of mass media entertainment for conversation, high mobility, rapid changes in jobs, etc., have made it increasingly difficult for people to establish and maintain intimate relationships.

Stability (Permanence and Continuity). Whole neighborhoods and towns change in character nowadays with remarkable swiftness. Even if the buildings themselves remain in place, the residents may change completely.

Security of Status. A person's position in society is increasingly uncertain due to marital instability and rapidly changing technology.

Modular life insurance products satisfy the scarcities of "time" and "wisdom". Insurance, in general, strengthens individual "stability" and "security". But although product technology fills several of these scarcities, the industry has a long way to go to communicate its ideas effectively to the market.

An interesting and challenging research project would be to express the life insurance product in terms of its psychic Value Added—namely its value in addressing the public's need for more time, more recognition, more wisdom, more influence, more intimacy, more stability, and more security. It could present an interesting illustration of the application of utility theory.

4. FUTURE DISTRIBUTION OF LIFE INSURANCE: EFFECTS OF UNIVERSAL LIFE

The business of life insurance is very vulnerable to competition both from inside and out. To begin with, ideas cannot by copyrighted so our basic technology is in the public domain. In addition, the business is moving away from insurance and towards assurance, and with that movement the levels of risk and the levels of capital required are diminishing. Moreover, the life insurance business is not dependent upon raw materials, energy, transportation and the other normal inhibitors of the commodity industry. These considerations should cause competition to increase. With today's technology, it is possible to create virtually unlimited product at point of sale through telecommunications. Consequently the "manufacturing" capacity of any one of the top 100 companies is probably sufficient to produce ALL the product required in North America. What this means is that the balance is tipping away from Manufacturing and towards Distribution. The strategic necessity is that companies must find secure lines of distribution or face annihilation.

I begin my speculation about future distribution by asking the question—What is distribution? By its general definition, distribution is the process of following a line of communication from the supplier to the customer. *A priori,* there is no "best" line of communication or distribution. In a market economy, the line that should be followed is the one that is most efficient. The life insurance industry has built its own lines of communication to the market but, in times of rapid change and high inflation, the lines of communciation that we have developed for permanent, ordinary life insurance have become increasingly costly to maintain or to extend and, therefore, are being supplemented or supplanted by other distribution systems. It is a reasonable speculation that, in time, the means of distribution will become more important than the means of manufacturing and that distribution lines will broaden to include "alien" systems, so that the balance of distribution power will swing away from traditional life insurance companies using conventional distribution systems towards non-traditional systems. Today, only slightly more than 50% of the ordinary life insurance sold in the United States is the result of distribution where the initiating party is the conventional agent or the conventional broker. If we were to look ahead a decade or two we might foresee an environment in which the manufacturing process is an appendage of distribution as compared to the present circumstance where distribution is an appendage of insurance manufacturing.

How does Universal Life fit into this picture? Some of the answers fall into place when we make simple comparisons between the strategies employed in marketing and the strategies employed in warfare. In the case of warfare, technological changes in weaponry and the delivery systems for new weaponry have gone hand in hand. The era of the foot soldier (delivery system) and spear (technology) gave way to the archer and bow (delivery system) and arrow (technology), and so on. At each stage of technological change there was a fundamental change in the strategy of warfare. There was no thought, for example, to employ a bow (old delivery system) to shoot a musket ball (new technology). The message that we get from the study of warfare is that the delivery system has to be appropriate to the technology. In our terms, distribution has to be appropriate to the product and vice versa. Using McLuhan's terms, the distribution system is the "content" of the product. Based upon historical record, new technology is almost invariably inappropriate to old delivery systems. Sometimes the new technology destroys the old delivery system—an example is the airplane and the battleship. If the existing delivery system of life insurance fails to adapt to new product and communications technology, it too may face attrition or extinction.

These illustrations suggest that emerging technology has opened up remarkable opportunities, either for employing existing distribution systems in new ways or for introducing completely new distribution systems.

It is highly relevant to consider the change that has taken place in communications. In the November 10, 1980 issue of *Business Week* magazine it is predicted that the percentage of homes in North America with computers or terminals having access to remote data bases will increase from 10% in the period 1980–1985 to one-third of the homes in 1985–90 to "most" of the homes in 1990–2000. This change in communications or distribution, could have enormously far-reaching implications for life insurance products and marketing. The skeptics will still ask, "Why bother about distribution through telecommunications when the life insurance industry is doing a very effective job of distribution using conventional systems?" The answer is that it is entirely possible that most consumer buying decisions in the 1990's will be made at home, without significant human intervention. If life insurance of the future is going to be marketed by pre-selling or by selling on the basis of short audio-visual presentations, then the life insurance product of the future will have to be both simple and coherent. This is both the medium and the message of Universal Life.

Those who search for new forms of life using the conventional Value Added approach of product stereotypes are trying to prolong the life of outmoded technology or outmoded delivery systems. The lesson from the strategies of warfare is that new delivery systems require new technology.

5. FUTURE ORGANIZATIONAL STRUCTURE: EFFECTS OF UNIVERSAL LIFE

The logical inference is that disassembly of product will accelerate disassembly of corporations. Corporate disassembly could either be real or simulated.

Many of you are aware that real disassembly has already occurred to a considerable extent. We have seen that distribution is becoming separated from manufacturing; many life insurance companies have set up separate marketing subsidiaries. But we have also seen the creation of separate subsidiaries for investment purposes, for claims payment, for administration, for actuarial consulting, for risk and reinsurance. The changes that have already taken place are not the result of Universal Life. On examination, these changes seem to be the result of product disassembly that has been taking place in Group operations for the last 30 years or more. To draw a parallel, the same kind of corporate disassembly should take place in life insurance companies whose principal business is the Individual or Ordinary line.

The reason for this conclusion is that, with fragmentation of product, for purely competitive reasons it is vital that costs be known by product fragment rather than in total (by product package, or stereotype). True Universal Life contemplates the assembly predicated to a considerable extent upon the unit costs of these manufacturers. Does this mean that a life insurance company that sells only Individual life insurance should undergo corporate disassembly? The answer is "No" for at least two reasons.

The first reason is that product disassembly can be simulated. In fact, the so-called functional form of corporate organization (as opposed to the federal decentralization or "line" form) has exactly that effect. Functional structure is a means of systematically separating all of the product fragments. Thus, if you look at an income statement, the distribution function is represented by premiums, the investment function by investment gains and losses, the underwriting function by claims payments, and so on. However, the same separation can be achieved by companies that operate on the basis of a product line, or a regional structure or a market structure, merely through simulation. In fact, most "line" companies conduct their financial activities in two ways—by "bottom line" accounting which is a method of accounting where the totals are established by "column"— and the "budget" accounting method, where the totals are established by "row" or function.

The second reason is that, increasingly, organizational structure is symbolic. If a company is growth-oriented, it wants to convey that objective in its formal organizational structure. So, in the usual case, organizational structure according to market will be used, with all other functions subordinated in the formal organizational chart. However, with the advances being made in information technology, increasing emphasis will be placed on what is termed the "computer-integrated-company" where decisions will be made through computer matching of those disciplines or functions which have decision-making responsibility for the problem at hand. Consequently, there should be ample freedom to design corporate structure for purely symbolic purposes, uninhibited by the real organizational processes or the real objectives of the company.

It would be logical to expect that product disassembly would mean corporate disassembly. As I have suggested, that will almost certainly take place, but the issue is whether this dissassembly will be real or simulated. On balance, consistent with Marshall McLuhan's maxim, corporate structure is the "content" of our product technology. So it should derive its form from product. However, the magic of information technology could well override formal organizational structure and make them extraneous.

6. CONCLUSIONS

In conclusion, my topic The Search For New Forms of Life has led me in a number of different directions. I have made some predictions about the effect of modular product on our profession and, in particular, upon the life insurance business. To a considerable extent, my remarks rely upon the thesis that one of the main purposes of universities is to present challenges. You might conclude that I have accomplished this, but only by satisfying the aphorism attributed to Samuel Butler—"Life is the art of drawing sufficient conclusions from insufficient evidence".

I admit to you that from the beginning of my actuarial career I have been a compulsive simplifier. However, I always try to observe the admonition of Albert Einstein: "Everything should be made as simple as possible, but not simpler."

REFERENCES

Business Week, November 10, 1980.

Canadian Journal of Life Insurance (1982), "Who invented Universal Life?"

Dinney, G. R. (1971), "A descent into the maelstrom of the insurance future". Canadian Institute of Actuaries.

Dinney, G. R. (1978), "A day in the life of your future". Life Underwriter Associates of Canada.

Dinney, G. R. (1982a), "Life insurance as a game". *ARCH, 1982.1*, 251–279.

Dinney, G. R. (1982b), "Product innovation in individual life insurance". Society of Actuaries Seminars.

Dinney, G. R. (1986), "Universal Life unde vadis, Quo Vadis". *Broker World.*

The Actuary (1981), "Universal Life—who carries this instrument?"

The Pacific Insurance Conference (1975), "The Universal Life Insurance Policy". James C. H. Anderson—footnote acknowledgment to G. R. Dinney.

THE UNIVERSITY OF WESTERN ONTARIO SERIES IN PHILOSOPHY OF SCIENCE

A Series of Books in Philosophy of Science, Methodology, Epistemology, Logic, History of Science, and Related Fields

1. J. Leach, R. Butts, and G. Pearche (eds.), *Science, Decision and Value.* 1973, vii + 219 pp.
2. C. A. Hooker (ed.), *Contemporary Research in the Foundations and Philosophy of Quantum Theory.* 1973, xx + 385 pp.
3. J. Bub, *The Interpretation of Quantum Mechanics.* 1974, ix + 155 pp.
4. D. Hockney, W. Harper, and B. Freed (eds.), *Contemporary Research in Philosophical Logic and Linguistic Semantics.* 1975, vii + 332 pp.
5. C. A. Hooker (ed.), *The Logico-Algebraic Approach to Quantum Mechanics.* 1975, xv + 607 pp.
6. W. L. Harper and C. A. Hooker (eds.), *Foundations of Probability Theory, Statistical Inference, and Statistical Theories of Science.* 3 Volumes. Vol. I: *Foundations and Philosophy of Epistemic Applications of Probability Theory.* 1976, xi + 308 pp. Vol. II: *Foundations and Philosophy of Statistical Inference.* 1976, xi + 455 pp. Vol. III: *Foundations and Philosophy of Statistical Theories in the Physical Sciences.* 1976, xii + 241 pp.
7. C. A. Hooker (ed.), *Physical Theory as Logico-Operational Structure.* 1979, xvii + 334 pp.
8. J. M. Nicholas (ed.), *Images, Perception, and Knowledge.* 1977, ix + 309 pp.
9. R. E. Butts and J. Hintikka (eds.), *Logic, Foundations of Mathematics, and Computability Theory.* 1977, x + 406 pp.
10. R. E. Butts and J. Hintikka (eds.), *Foundational Problems in the Special Sciences.* 1977, x + 427 pp.
11. R. E. Butts and J. Hintikka (eds.), *Basic Problems in Methodology and Linguistics.* 1977, x + 321 pp.
12. R. E. Butts and J. Hintikka (eds.), *Historical and Philosophical Dimensions of Logic, Methodology and Philosophy of Science.* 1977, x + 336 pp.
13. C. A. Hooker (ed.), *Foundations and Applications of Decision Theory.* 2 volumes. Vol. I: *Theoretical Foundations.* 1978, xxiii + 442 pp. Vol. II: *Epistemic and Social Applications.* 1978, xxiii + 206 pp.
14. R. E. Butts and J. C. Pitt (eds.), *New Perspectives on Galileo.* 1978, xvi + 262 pp.
15. W. L. Harper, R. Stalnaker, and G. Pearce (eds.), *Ifs. Conditionals, Belief, Decision, Chance, and Time.* 1980, ix + 345 pp.
16. J. C. Pitt (ed.), *Philosophy in Economics.* 1981, vii + 210 pp.
17. Michael Ruse, *Is Science Sexist?* 1981, xix + 299 pp.

18. Nicholas Rescher, *Leibniz's Metaphysics of Nature.* 1981, xiv + 126 pp.
19. Larry Laudan, *Science and Hypothesis.* 1981, x + 258 pp.
20. William R. Shea, *Nature Mathematized.* Vol. I, 1983, xiii + 325 pp.
21. Michael Ruse, *Nature Animated.* Vol. II, 1983, xiii + 274 pp.
22. William R. Shea (ed.), *Otto Hahn and the Rise of Nuclear Physics.* 1983, x + 252 pp.
23. H. F. Cohen, *Quantifying Music.* 1984, xvii + 308 pp.
24. Robert E. Butts, *Kant and the Double Government Methodology.* 1984, xvi + 339 pp.
25. James Robert Brown (ed.), *Scientific Rationality: The Sociological Turn.* 1984, viii + 330 pp.
26. Fred Wilson, *Explanation, Causation and Deduction.* 1985, xviii + 385 pp.
27. Joseph C. Pitt (ed.), *Change and Progress in Modern Science.* 1985, viii + 398 pp.
28. Henry B. Hollinger and Michael John Zenzen, *The Nature of Irreversibility.* 1985, xi + 340 pp.
29. Kathleen Okruhlik and James Robert Brown (eds.), *The Natural Philosophy of Leibniz.* 1985, viii + 342 pp.
30. Graham Oddie, *Likeness to Truth.* 1986, xv + 218 pp.
31. Fred Wilson, *Laws and Other Worlds.* 1986, xv + 328 pp.
32. John Earman, *A Primer on Determinism.* 1986, xiv + 273 pp.
33. Robert E. Butts (ed.), *Kant's Philosophy of Physical Science.* 1986, xii + 363 pp.
34. Ian B. MacNeill and Gary J. Umphrey (eds.), Vol. I, *Applied Probability, Stochastic Processes, and Sampling Theory.* 1987, xxv + 329 pp.
35. Ian B. MacNeill and Gary J. Umphrey (eds.), Vol. II, *Foundations of Statistical Inference.* 1987, xvii + 287 pp.
36. Ian B. MacNeill and Gary J. Umphrey (eds.), Vol. III, *Time Series and Econometric Modelling.* 1987, xix + 394 pp.
37. Ian B. MacNeill and Gary J. Umphrey (eds.), Vol. IV, *Stochastic Hydrology.* 1987, xv + 225 pp.
38. Ian B. MacNeill and Gary J. Umphrey (eds.), Vol. V, *Biostatistics.* 1987, xvi + 283 pp.
39. Ian B. MacNeill and Gary J. Umphrey (eds.), Vol. VI, *Actuarial Science.* 1987, xvi + 250 pp.

GPSR Compliance

The European Union's (EU) General Product Safety Regulation (GPSR) is a set of rules that requires consumer products to be safe and our obligations to ensure this.

If you have any concerns about our products, you can contact us on ProductSafety@springernature.com

In case Publisher is established outside the EU, the EU authorized representative is:

Springer Nature Customer Service Center GmbH
Europaplatz 3
69115 Heidelberg, Germany

Batch number: 10607222

Printed by Printforce, the Netherlands